Doctor Who and Science

Doctor Who and Science
Essays on Ideas, Identities and Ideologies in the Series

Edited by MARCUS K. HARMES
and LINDY A. ORTHIA

McFarland & Company, Inc., Publishers
Jefferson, North Carolina

LIBRARY OF CONGRESS CATALOGUING-IN-PUBLICATION DATA

Names: Harmes, Marcus K., editor. | Orthia, Lindy A., 1972– editor.
Title: Doctor Who and science : essays on ideas, identities and ideologies in the series / edited by Marcus K. Harmes and Lindy A. Orthia.
Description: Jefferson : McFarland & Company, Inc., Publishers, 2021 | Includes bibliographical references and index.
Identifiers: LCCN 2020056705 | ISBN 9781476681122 (paperback : acid free paper) ∞
ISBN 9781476642000 (ebook)
Subjects: LCSH: Doctor Who (Television program : 1963–1989) | Doctor Who (Television program : 2005–) | Science on television. | Science fiction television programs—Great Britain—History and criticism.
Classification: LCC PN1992.77.D6273 D6225 2021 | DDC 791.45/72—dc23
LC record available at https://lccn.loc.gov/2020056705

BRITISH LIBRARY CATALOGUING DATA ARE AVAILABLE

ISBN (print) 978-1-4766-8112-2
ISBN (ebook) 978-1-4766-4200-0

© 2021 Marcus K. Harmes and Lindy A. Orthia. All rights reserved

No part of this book may be reproduced or transmitted in any form or by any means, electronic or mechanical, including photocopying or recording, or by any information storage and retrieval system, without permission in writing from the publisher.

Front cover image © 2021 Shelby Allison/Shutterstock

Printed in the United States of America

*McFarland & Company, Inc., Publishers
Box 611, Jefferson, North Carolina 28640
www.mcfarlandpub.com*

Acknowledgments

The editors would like to thank the contributors for their timely preparation of high-quality essays and revisions. Some contributors agreed to review other essays and we are grateful for their useful feedback. Thanks also to those who submitted an abstract for consideration but were not offered the opportunity to write an essay because of space constraints: we look forward to seeing your work published elsewhere. Thanks to McFarland, especially Layla Milholen, for rapid approval of our project and continuing support. On behalf of contributors we would also like to thank the universities and other institutions that supported all of us to complete this work.

Table of Contents

Acknowledgments   v

Timeline and Terminology for Doctor Who's *Doctors and Eras*   1

Introduction to Doctor Who *and Science*
    Lindy A. Orthia *and* Marcus K. Harmes   3

Who's Planet Looks Like Home?
    J.J. Eldridge   18

Who's Moon
    Elizabeth R. Stanway   33

$E=mc^3$: *Doctor Who* and Energy
    Marcus K. Harmes   48

Translation by TARDIS: Exploring the Science Behind Multilingual Communication in *Doctor Who*
    Mark Halley *and* Lynne Bowker   62

"I don't want to go": How Does Regeneration Work in *Doctor Who*?
    Natalie Ring   78

Did the Doctor Change Sex or Change Gender? Navigating the Sex and Gender Divide in *Doctor Who*
    Mike Stack   94

Candyfloss, Lego and Hope: What Sort of Scientist Is Jodie Whittaker's Doctor?
    Lindy A. Orthia *and* Vanessa de Kauwe   110

The Mad Scientist Wore Prada: Female Frankensteins in the Universe of *Doctor Who*
    Kristine Larsen   127

Maxtible's Mirrors: Victorian Science in Classic-Era *Doctor Who*
    Marcus K. Harmes *and* Richard Scully   142

The Victorians Sleeping in Our Minds: Victorian Scientific Enquiry in Old and New Series *Doctor Who*
    Catriona Mills   157

Doctor Who and the Dinosaurs: Spectacle, Monstrosity, Melodrama and Ideology in Dinosaur Mediations
    Ross Garner   173

The Use and Abuse of Scientific Writing in *Doctor Who*'s Epistolary Paratexts
    Tonguç İbrahim Sezen   190

The Science of *Doctor Who*
    Mark Erickson   205

Concluding Remarks: Science in Twenties Doctor Who
    Lindy A. Orthia *and* Marcus K. Harmes   221

About the Contributors   227

Index   231

Timeline and Terminology for *Doctor Who*'s Doctors and Eras

Numerous actors have played *Doctor Who*'s main character, the Doctor, since the program's inception in 1963.

Classic Series, 1963–1989
William Hartnell 1963–66
Patrick Troughton 1966–69
Jon Pertwee 1970–74
Tom Baker 1974–81
Peter Davison 1982–84
Colin Baker 1984–86
Sylvester McCoy 1987–89

US-UK Television Movie, 1996
Paul McGann

New Series, 2005–present
Christopher Eccleston 2005
David Tennant 2005–10
Matt Smith 2010–13
Peter Capaldi 2013–17
Jodie Whittaker 2017–present

Other Doctors
John Hurt 2013
Jo Martin 2020

The thirteen actors from the classic series, television movie and new series lists are also often known by their number in that sequence, for example William Hartnell is known as "the First Doctor" and Jodie Whittaker as "the Thirteenth Doctor." These terms are used by some authors in this book

2   Timeline and Terminology

and correspond to series leads. The leads also mark major eras of *Doctor Who*, for example *Doctor Who* commentators might speak of "the Eleventh Doctor era" or "the Pertwee years." However, we note the sequence is complicated both by the non-lead appearances of John Hurt and Jo Martin as the Doctor, and by other actors playing the different incarnations, in particular Richard Hurndall (1983) and David Bradley (2017) portraying the First Doctor after William Hartnell's death.

Introduction to *Doctor Who* and Science

LINDY A. ORTHIA *and* MARCUS K. HARMES

The actor William Hartnell, who played the first incarnation of the Doctor, was not especially interested in science, but he reminisced about a singular impact playing the role had for him: his viewers began to ask him for help with science. He said he would "get letters from boys swotting for O-levels asking complicated questions about time-ratio and the TARDIS. The Doctor might have been able to answer them—I'm afraid I couldn't!" On another occasion there was another request he could not comply with, when scientists at the University of Nottingham invited him to deliver a course of lectures (Carney 2013: 165; IMDb 2019). These requests may have been misdirected but are suggestive, indicating that for some people *Doctor Who* was received as a source of scientific instruction. Hartnell's anecdotes intersect with one of *Doctor Who*'s prevailing foundation stories, namely that television executive Sydney Newman intended the BBC's new drama series to teach children about science and history as they watched the adventures of the Doctor and his companions (Burk and Smith 2012; Cooke 2003; O'Brien 2000; Robb 2009). The intention speaks to the BBC's Reithian objectives to educate as well as entertain (even though Newman was himself a product of other broadcasting cultures, Canadian and British independent), and is a meaningful starting point for a collection on science in *Doctor Who*.

However, why just science and history? Is the foundation story even correct that the new teatime serial was to be educational? Did the idea ever go further than a suggestion from Sydney Newman and ever actually manifest on screen? Is it possible to learn any science from *Doctor Who*? To consider these points in order, the suggestion that *Doctor Who* was going to teach school children about science and history can seem like an explanation retrofitted to explain scripting and casting choices, in that the original characters joining the Doctor in the TARDIS were a male science teacher, Ian Chesterton, and a female history teacher, Barbara Wright. From time to time, Barbara authoritatively foregrounded her historical knowledge; it is intrinsic to the plotting of "The Aztecs" (1964) and in "The Dalek Invasion of Earth" (1964) gives her mouthfuls of semi-plausible content to bamboozle the Daleks into thinking that Hannibal, General Lee and American revolutionaries were all about to attack. Ian's science knowledge though occupied a more contested space. The presence of two aliens from an advanced species, the Doctor and Susan, undercut and even rendered irrelevant Ian's scientific knowledge, as the Doctor and Susan carried the dramatic function of scientific exposition. Ian's first appearance shows him complaining about being shown up in class by a precociously intelligent schoolgirl, in reality the alien Susan (Harmes 2016).

But why just science and history? The curricula of both secondary modern and grammar schools in 1960s Britain ranged far more widely than these disciplines. But there is no suggestion in contemporary paperwork or in the many later histories of *Doctor Who*'s creation that the series could also have taught mathematics, geography, home economics, foreign languages or English, or anything else being taught to British school children. Why was there a perceived need for an educational adventure series and did such things exist on screen? *Doctor Who*'s creative crucible and its broadcasting context both show that school children were in fact well catered to if they wished to watch television programs about science. On the day of *Doctor Who*'s first transmission, November 23, 1963, school children had already been able to watch a great deal of science. Sir Hermann Bondi, the Cambridge mathematician and colleague of Fred Hoyle and Thomas Gold, presented episode 8 of the televised lecture series $E=mc^2$: *Thinking Relativity Through* on the BBC Television Service, with models made by the same BBC Visual Effects Department then busy for Verity Lambert, *Doctor Who*'s producer, creating prehistoric caves and the petrified forests of Skaro. Bondi's presentation was part of BBC Television's thorough engagement with science on the small screen. In September 1963, the BBC had commenced a series on *Men of Science*, starting with Sir Humphry Davy. Science broadcasting continued apace in 1964 with *Engineering Science*, *Science and Life*, *Indoors and Out: Science in the Garden* and *Discovering Science* appearing that year, among others. The broadcasts brought onto the screen and made available a number of major thinkers, including the neuroscientist Gerd Sommerhoff on *Discovering Science*, presenting on "What Is Water" in March 1963. The presenters were largely all male, reflective of a gender disparity which continues to inform discourses of opportunities, recognition and careers in STEMM fields (science, technology, engineering, mathematics and medicine), and indeed making the science teacher male, and the history teacher female is reflective of the male-dominated way science was then appearing on screen.

There was diversity though in subject matter. Anatomy, chemistry, botany, biology, applied science, science in the home, geology, optics, magnetism and more featured in television broadcasts during daytime hours and intended for broadcast in schools, raising the question of whether British school children of the early 1960s may necessarily have wished to go home and watch yet more educational science content in a science fiction serial. Would watching children necessarily have learned any science from *Doctor Who*? In the first episode, "An Unearthly Child" (1963), the exchanges between Ian and the Doctor and Susan immediately complicate the possibility that the series could be educational. The episode had already shown flashbacks in which Susan demeaned Ian's science lessons at the Coal Hill School as childish and pointless. Once Ian enters the TARDIS and meets the Doctor, the Doctor derides Ian as childish, ignorant, and arrogant, and tells him he would be unable to understand the science and technology inside the ship. Viewers themselves learn that the science the Doctor uses comes from "another time, another world."

Nonetheless, from time to time, early *Doctor Who* could provide small-scale lessons in science. In "Marco Polo" (1964), when the travelers are stranded in medieval China without drinking water, a simple scientific process becomes central to the plot. Ian and the Doctor explain to the Italian merchant Marco Polo that they had not been hoarding water in their "caravan" (the TARDIS), instead "The outside of the caravan cooled, but the inside stayed warm, and so moisture formed on the inside. It's condensation." "Planet of Giants" (1964) imparts different lessons. Two years earlier in the United States, Rachel Carson had written *Silent Spring*. This *Doctor Who* story responds to the themes of her environmental study when Barbara, in a garden surrounded by dead insects, asks, "surely it's wrong to kill bees and worms and things, isn't it?" and the Doctor explains their importance, as "Both are vital to the growth of things." These though are small moments, certainly not part of any systematically thought-through way of teaching science via drama. Set against them is another much-discussed aspect of *Doctor Who*'s creation, in that Verity Lambert chose against Sydney Newman's wishes to introduce the Daleks and take the series into the realm of science fiction rather than science education.

From time to time thereafter, particular production personnel gave attention to introducing putatively credible science into narratives. Dr. Kit Pedler, an ophthalmologist, contributed ideas and scripts to 1960s serials, as well as contributing to the creation of the Cybermen. However, Pedler's expertise in eye surgery was removed from the science fiction in his scripts, and his ideas were arguably products of his personal preoccupations and anxieties. In the next decade, producer Barry Letts brought scientific discourse into the production office, notably as a reader of *New Scientist* (Hearn 2013). In 1980, incoming script editor Christopher H. Bidmead purportedly sought to introduce "hard" science into stories (Chapman 2006). His intentions and his success in doing so are subject to debate. What is notable, however, is that having science in the scripts did not depend on self-consciously scientific writers and script editors. Douglas Adams, Bidmead's predecessor as both a scriptwriter and a script editor and who oversaw a period frequently recalled as more frivolous than Bidmead's, read English literature at Cambridge, not one of the sciences. Yet he successfully incorporated science into plots and dialogue. In "The Pirate Planet" (1978) for example, the Doctor and Romana's escape from pursuing guards along an anti-inertia corridor uses the laws of physics in the plotting. "You know, I think the conservation of momentum is a very important law in physics," remarks the Doctor, as he switches off the inertia neutralizer. When the guards tumble out of the corridor and hit a wall, having conserved their momentum, the Doctor quips it is "Newton's revenge."

The notion of teaching science via drama receives critique in David Layton's study of philosophical ideas in the series (2014). As he points out, *Doctor Who*, far from teaching science, can actually propagate wildly inaccurate ideas (he gives as an example the purported science about the Moon in "The Silurians," 1970). Even if the content may have been accurate, Layton also makes the valid point that the best science fiction can teach is at the macro level, such as introducing science and reason as major concepts. As we will see, *Doctor Who* does deliver on that level, especially when the Doctor educates less "enlightened" companions to have faith in science. The story that Sydney Newman created *Doctor Who* to teach science has, however, lingered, being promoted in histories of the program and in interviews with its creators. Watching early *Doctor Who* provides very little instruction and the schoolchildren hoping William Hartnell could help them study for their exams were going to be disappointed, and were unlikely to learn much from his show. However, the story is important for establishing one critical point: that from the outset *Doctor Who* was going to be about science.

Scientists in Doctor Who

Whatever the original motivation for its creation, *Doctor Who* has engaged richly with science-related characters and themes during the many decades of its production. Most obviously, the Doctor identified strongly as a scientist during the Hartnell era, describing his adventures in time and space as a "scientific expedition" in "Galaxy Four" (1966). Five incarnations who followed Hartnell (those of Patrick Troughton, Jon Pertwee, Tom Baker, Colin Baker and Jodie Whittaker) also described themselves as scientists, and some heavily promoted science as a belief system and problem-solving method. All of the Doctor's incarnations up to and including Whittaker's have extensively engaged in technical and laboratory-based work and have relied upon scientific knowledge during their adventures, enabling the program to be identified with science by many viewers. The strongest association may be in the first seasons of the Pertwee era, which saw the Doctor appointed as the official scientific advisor to the United Nations Intelligence Taskforce (UNIT). The appointment turned him into a science professional rather than a wandering and wondering explorer, now seen frequently in a laboratory. The third incarnation of the Doctor described himself as "every kind of scientist" ("Colony in Space," 1971), staking a claim to multiple realms of expertise. This association with UNIT lingered to an extent into the Tom Baker, Peter Davison and Sylvester McCoy eras, and then recurred in the new series with the Doctor again providing scientific support to UNIT. Yet these portrayals have varied considerably, and the type of scientist (or science-knowledgeable person) the Doctor embodies has varied across the series. Scholars' different categorizations of the character demonstrate this. For example Roslyn Haynes (2003) categorized the Doctor as a "scientist adventurer" when discussing her seven scientist stereotypes in Western fiction, Lindy A. Orthia (2010) in contrast positioned the Doctor as one of Haynes' "noble scientists," Robert Jones (1997) classified the character (especially Pertwee's incarnation) within the boffin stereotype common to post-war British films, and Alec Charles (2013: 83) categorized the "eccentric," "brilliant" and "emotionally stunted" Doctor as a Baudelairean *flâneur*.

Besides the Doctors themselves, many of the Doctor's traveling companions also trained in a STEMM field. These include science teacher Ian, math teachers Danny Pink and Brigadier Lethbridge-Stewart, mathematicians Zoe Heriot and Adric, research physicist Liz Shaw (who also had qualifications in medicine and other fields), medical doctors Martha Jones, Harry Sullivan and Grace Holloway, nurse Rory Williams, biochemist Nyssa, computer programmer Mel, archaeologists River Song and Bernice Summerfield, and science students Peri Brown (botany) and Bill Potts (physics). The scientist side of these characters frequently fell by the wayside, with narratological pressures—not to mention ambient sexism, racism and class prejudice—often pushing companions toward damsel or fighter roles, leaving the Doctor to fulfill the scientific expert function solo (Jowett 2014; Orthia 2010; Powers 2016; Tulloch and Alvarado 1983). Perhaps most notoriously, the Doctor and everyone else addressed the multiply qualified

Dr. Liz Shaw as "Miss," and her expertise was ignored and sidelined after her first serial, with her role as UNIT's scientific advisor usurped by the Doctor. In addition, cameras paid unnecessary attention to her miniskirted legs in "The Silurians," as they similarly did to Zoe's form-fitting catsuit in "The Mind Robber" (1968) and Peri's bikini in "Planet of Fire" (1984). The Doctor also mocked the fact that former soldiers Danny and the Brigadier were math teachers, in examples of subtle racist and classist attitudes disguised as anti-militarism. But these unfortunate characterizations were not always the last word for scientist companions. Some, such as Nyssa, Martha and River, kept their brains, their skills and their role-modeling capacity to the end (Coppa 2010; Larsen 2018; Orthia 2010; Patrick 2017). Sometimes plotting and intelligence effectively synchronized, including Nyssa's biochemical knowledge, which provided an elegantly plotted reason for her departure in "Terminus" (1983).

Other companions were self-taught in scientific areas or trained through experience, such as Leela, whose knowledge of anatomy derived from her warrior training, Ace, who taught herself to make explosives, Mickey Smith, a self-trained computer hacker, and Wilfred Mott, an amateur astronomer. Still others were technologically skilled or scientifically knowledgeable by virtue of being aliens, robots or humans from the future. Notable examples are the Doctor's fellow Time Lord Romana, a high achiever at the Gallifreyan Academy who among other things constructed a sonic screwdriver that eclipsed the Doctor's, and future human Jack Harkness, who led the Earth-defending Torchwood Institute with his extensive knowledge of alien technologies in the spin-off series *Torchwood* (2006–11). Scientific knowledge and skills—or the lack of them—have not always been a presence on the TARDIS.

Emphasizing the centrality of science to some eras of *Doctor Who*, at times the Doctor explicitly coached companions they (or the show's production crew) perceived to be insufficiently knowledgeable in science. Famously, Pertwee's Doctor tried to initiate his companion Jo Grant, who had failed her O level in science, into scientific ways of thinking. He did not relinquish this ambition when she left to marry a Nobel Laureate biologist, expressing hope that her husband "might even be able to turn [her] into a scientist" ("The Green Death," 1973). This was a relationship dynamic based on an imbalance of knowledge and expertise, in which the Doctor's intelligence was offset by his assistant's relative ignorance. Before viewers even see Jo in her first serial ("Terror of the Autons," 1971), we learn of her ostensibly lowly intellectual function, whereby the Brigadier said, "What you need, Doctor, as Miss Shaw herself so often remarked, is someone to pass you your test tubes and to tell you how brilliant you are. Miss Grant will fulfil that function admirably." Tom Baker's Doctor had rather more success (from a scientistic perspective) in patronizingly preaching science to Leela, whom he problematically persisted in calling a "savage" in a reproduction of stagist Eurocentric values (Orthia 2013). Leela's narrative arc seemed to demonstrate the rightness of the Doctor's goal, when she proselytized to an Edwardian woman who placed faith in astrology, "I too used to believe in magic. But the Doctor has taught me about science. It is better to believe in science" ("Horror of Fang Rock," 1977). In a less obvious example of scientific didacticism, former shop assistant Rose Tyler learned enough about aliens during her travels with Christopher Eccleston's Doctor and David Tennant's to find work as an advisor for Torchwood, an achievement of which the Doctor approved. Some other non-scientist companions had similar learning trajectories to Rose, notably journalist Sarah Jane Smith who became another of Earth's alien-savvy defenders in her spin-off series *The Sarah Jane Adventures* (2007–11).

Aside from the Doctors and companions, *Doctor Who* has featured literally hundreds of scientist characters as enemies or allies over the years. Some of the Doctor's recurring enemies are scientists, including the Master, a power-hungry dabbler in many sciences; the Rani, a brilliant and amoral biochemist and bioengineer; Omega, the exiled stellar engineer whose work powered the Time Lords' time travel experiments; and Davros, the genetics and cybernetics genius who created the Daleks. Beyond those regulars, one study counted 222 human or human-looking scientist characters who had a prominent role in one or two televised serials during the program's first 50 years (Orthia and Morgain 2016). This number did not include the many non-human-looking scientists featured in the program, scientists with only minor roles in their stories, nor characters in *Doctor Who* novels, audio adventures or spin-offs; a full count of all of these would send this number soaring. It is notable that while the ratio of female to male scientists in *Doctor Who* has been far from 50:50, it has become more balanced over time, and on average, female scientists are characterized as equally scientifically credible as their male counterparts throughout the series (Orthia and Morgain 2016). The program has also positively depicted some trans and queer characters with scientific expertise, such as the pansexual and panscientific Jack Harkness, archaeologist and implied lesbian Professor Emilia Rumford from "The Stones of Blood" (1978), and trans mathematician Eleanor Blake from audio adventure "The Jabari Countdown" (2018). Arguably, the Doctor may also be counted among such characters, having regenerated from male to female at the end of Peter Capaldi's tenure in the role, and having implied romantic associations with women and men while male. However, the show has also frequently equated gender non-conformity and non-masculinist traits with scientific incompetence, reflecting a masculinist scientific culture in the real world (Orthia and Morgain 2016). Whatever its take, it cannot be denied that *Doctor Who* has presented viewers with a whole lot of scientist characters to think about, argue with or learn from.

Science Issues, Ideologies and Ideas in Doctor Who

Scientists are common in *Doctor Who* because its serials frequently engage with science-related themes or threats, and scientist characters put a human face on the ideological stances involved. Scientific institutions, too, feature as settings for *Doctor Who*, enabling an exploration of science's impacts and meanings for society. In particular, *Doctor Who* commonly depicts ethical dilemmas in science. It would take too many words to name every scientific issue *Doctor Who* has dealt with in its nearly 60 year history to date, but the list includes: weapons technologies; artificial intelligence, robotics and cybernetics; genetic modification, cloning and eugenics; animal experimentation; pollution and environmentalism; nuclear power and other energy sources; drug addiction and psychological control (Dubois 2015; Larsen 2013; Orthia 2010, 2011b). Its position on these topics has varied—sometimes romantically criticizing scientific hubris, other times valorizing technological innovation—though it has generally opposed technologies that are cruel or cause environmental destruction. It has also repeatedly told stories dealing with the relationships between science and colonialism, and science and religion or superstition, and science has played a significant role in maintaining or overthrowing political regimes in *Doctor Who* serials (de Kauwe and Orthia 2018; Fiske 1984; Johnson 2013; Morgain 2013; Orthia 2011a, 2013; Vohlidka 2013). Again, its engagement with these topics has sometimes been contradictory; for example, it has at times explicitly opposed the use of science as a colonialist tool, while employing scientific discourses that bolster the Eurocentric ideologies behind real world colonialism. It has also frequently "disproved" superstitious and religious beliefs with rationalist explanations, while elsewhere presenting religious beliefs in a favorable light. Despite the show criticizing specific instances of scientific hubris or unethical action, scholars tend to agree that *Doctor Who*'s ethos is generally pro-science (in particular, pro-Western science), and is consistent with Western Enlightenment values (Fiske 1984; Orthia 2010; Tulloch and Jenkins 1995). Proponents of this argument include Anne Cranny-Francis (2009: 124), who distinctively concluded that the structure of *Doctor Who*'s famous electronic theme music reflects Enlightenment values such as "the secular world of the bourgeoisie," "the notion of human progress and perfectability" and a "romantic view of science and technology."

Given the program's themes, John Tulloch and Manuel Alvarado (1983:

41) described *Doctor Who* as conforming to a "'soft' socio-cultural scientific speculation" model of science fiction, rather than a model of science fiction that prizes technical and factual accuracy. Many *Doctor Who* fans would laughingly agree with this evaluation, being far more likely to take *Doctor Who* seriously when thinking about science as a social endeavor than to use it as a science textbook, as per the short-lived (if ever actually achieved) ambition that *Doctor Who* would teach children science. Despite this, *Doctor Who* has been the subject of many "science of *Doctor Who*" texts designed to connect its scientific ideas and technologies with real world scientific concepts, such as time travel, robotics and sonic technologies (O'Keeffe 2017). These include numerous essays and articles (e.g., Larsen 2017; Oakes 2013; Richmond 2015), several exhibitions in science museums (e.g., Bell 2018), three books (Guerrier and Kukula 2015; Parsons 2006; White 2005), two documentaries (Cox 2013; O'Connor 2012), and at least one touring stage show (Williams 2014).

Its relevance to real world science—whether to the social, cultural, political, ideological and ethical aspects of science or to our understandings of space, time, aliens, biology, chemistry, mathematics and physics—is apparent when considering the program's impacts on its viewers. A 2015 survey of 575 science-interested *Doctor Who* fans from 37 countries revealed that the show had influenced the career choices of 9 percent of those surveyed, and the education choices of 13 percent, most of whom had chosen to work or study in a science field (Orthia 2019). In the same survey, about a quarter to a third of respondents said the show contributed to their ideas about science ethics (37 percent), their approaches to problem solving (36 percent), their vision of what the future should look like (31 percent), their ideas about the appropriate relationship between science and the rest of society (28 percent), and their ideas about the role played by science in human history (24 percent). More than a television show, *Doctor Who* is a significant participant in global conversations about the relevance of science to our daily lives.

The Essays

Despite the many works cited above that have discussed *Doctor Who* and science, there is still much to say on the topic, and we believe this book is long overdue. This is the first book to bring together essays that draw on the sciences in a "real science of *Doctor Who*" approach, essays that analyze *Doctor Who*'s depictions of science from literary, historical and sociological perspectives, and essays that critically examine popular and academic discourse on the topic of science and *Doctor Who* by drawing on fields such as media studies and cultural studies. Most of the essays in fact blur such disciplinary boundaries, creating a genuinely multi-disciplinary collection.

Two astronomical topics open the book. First up, J.J. Eldridge explores *Doctor Who*'s representations of alien planets from Mars to the gold planet Voga to the "impossible planet" orbiting a black hole, and more. They consider how these compare to scientists' understandings of the kinds of planets possible in the universe, dilemmas about terraforming and colonizing planets, and whether the show's planetary depictions have changed in concert with the evolving science. They also offer an enticing glimpse of *Doctor Who*'s impact on real-world astronomy and astrophysics.

Elizabeth R. Stanway follows with an examination of *Doctor Who*'s many depictions of Earth's Moon, including scientific, cultural and political frames. She raises questions about ownership and sovereignty of the Moon, uses of the Moon as a scientific space or a resource, and perceptions of the Moon as a potentially dangerous influence, comparing *Doctor Who*'s varied and sometimes fanciful Moon-based stories to real world science and political machinations. Finally, she reflects on waning interest in the Moon in the post–Apollo 1970s, in *Doctor Who* and beyond.

Continuing the theme of real world scientific developments shaping *Doctor Who* stories, Marcus K. Harmes tracks the show's apparent fascination with energy in serials produced between 1963 and 1980, many infused with energy-related plots, themes and science starting with the very first episode. Among the numerous scientist characters seeking unlimited energy sources, the apocalyptic radioactive landscapes, and alien technologies that suck the energy from bodies, he finds correspondences with contemporary fears of a nuclear disaster and anxieties about coal, oil and gas shortages in Britain.

Mark Halley and Lynne Bowker turn our attention from the geophysical sciences to linguistic engineering in their essay about the TARDIS translation circuit, which (sometimes) allows *Doctor Who*'s characters to communicate in what appears to be the same language. They draw on the real world science and history of machine translation and interpretation to speculate on how the TARDIS technology might work, what it would need to work as well as it does on screen, and the inconsistent (and perhaps inexplicable) circumstances when it doesn't work at all.

Moving from Time Lord technology to Time Lord bodies, Natalie Ring analyzes the process of regeneration that has kept *Doctor Who* on our screens for so many decades. She breaks it down scientifically, identifying the key traits of regeneration as portrayed in the show. She then looks to examples of plants and animals and human body cells that undergo regenerative processes of different kinds, to think about what might be going on in Gallifreyan biology that enables the Time Lords to cheat death this way.

Mike Stack offers a different perspective on regeneration, exploring whether Capaldi's Doctor changed sex or gender or something else when he regenerated into Whittaker's Doctor. He analyzes the program's depiction of Whittaker's Doctor, particularly in her first (post-regeneration) season, to interrogate what *Doctor Who* tells us about gender and sex with respect to gender theory, biological models of sex, and more. He concludes the Doctor is genderfluid, and highlights the liberatory prospects for how the show has navigated this in the Whittaker era thus far.

Lindy A. Orthia and Vanessa de Kauwe also examine Whittaker's Doctor, starting with the argument that the Doctor is widely perceived to be a scientist role model for viewers, and therefore Whittaker should be well placed to challenge sexist stereotypes about scientists. They ask what kind of scientist Whittaker's Doctor models, comparing her both to earlier Doctor incarnations and to historical, sociological and representational expectations of what a scientist is. They propose that the Doctor is best interpreted as a steward of scientific knowledge rather than a scientist.

Kristine Larsen shifts the focus to other female scientist characters in *Doctor Who*, specifically three villains: the Rani, the Master's female incarnation Missy, and Madame Kovarian, an important antagonist during Matt Smith's tenure as the Doctor. She examines the extent to which these characters are consistent with stereotypes of villainous scientists as mad or inhumane and sexist stereotypes that highlight women's attractiveness, fashion sense and motherliness (or the absence of these), to determine what kinds of messages these characters may send about women in science.

Marcus K. Harmes and Richard Scully in turn scrutinize one of the many male scientist characters *Doctor Who* depicts as both mad and inhumane: Theodore Maxtible from "The Evil of the Daleks" (1967). Drawing on the serial's rich characterization of this nineteenth-century gentleman amateur experimenter, they situate his science within the Victorian scientific milieu, with particular reference to electromagnetism pioneers Michael Faraday and James Clerk Maxwell. They argue Maxtible is just one example of the influence Victorian science exerted on twentieth-century science fiction, including other instances in *Doctor Who*.

Catriona Mills also explores the pervasive presence of Victorian scientific obsessions in *Doctor Who*, specifically dimensionality, archaeology and museum building. She interrogates representations of these themes in three *Doctor Who* serials: "Flatline" (2014) and its correspondences with texts like Edwin Abbott Abbott's 1884 book *Flatland* that explored the mathematics of two-, three- and four-dimensions; "Tomb of the Cybermen" (1967) and its links to imperialist archaeological endeavors and lost civilization romances; and "The Space Museum" (1965) as an example of the Victorian museum as self-conscious monument to (decaying) empire.

Moving from literary influences to cross-genre televisual mediation, Ross Garner examines depictions of dinosaurs and other "giant lizards" in *Doctor Who* to evaluate how the show has contributed to constructing and communicating ideas about dinosaurs and deep time. Drilling deeper than questions of scientific accuracy, the quality of dinosaur special effects, or a reading of dinosaurs as simply further entries in a long list of *Doctor Who* monsters, he argues these representations have multiple meanings for drama and melodrama, portrayals of the scientific gaze, and biodiversity-related ideologies.

Tonguç İbrahim Sezen completes the crossing into transmedia studies with an analysis of *Doctor Who*'s epistolary paratexts: the books and other publications that purport to show how alien technologies (such as the TARDIS or a Dalek) are built, and are presented as if they were technical manuals, scientific notebooks, or similar reference documents. He shows how these contributed to the *Doctor Who* franchise's growth and fans' attempts to create canonical coherence, through their allusions to real world science and technology and their gap-filling fabrication of scientific and technical information.

Finally, Mark Erickson examines three *Doctor Who* paratexts that use *Doctor Who* as a hook to engage audiences with science: two books and a televised lecture that were all (at least partly) entitled *The Science of Doctor Who*. He evaluates the varying extents to which these texts engaged with *Doctor Who* content, as well as identifying within each four tropes common to the popular science genre: triumphal histories of science, ideological assertions of science's superiority, uncritical collectivizations of many sciences into a singular "science," and male-centric bias.

Realistically, academics tend not to write about *Doctor Who* unless they are also fans or at least highly engaged viewers. The contributors to this volume, among us biologists, linguists and astrophysicists, historians, communication scholars and social scientists (and more), have applied the same scrupulous rigor in these essays as we would in other works. But underpinning our essays is a genuine enthusiasm—as well as humor about this sometimes ridiculous, sometimes brilliant television program—that aims to stimulate and provoke. We hope you enjoy reading the book as much as we all enjoyed writing it.

Before proceeding, we editors want to note that one of the questions we asked ourselves when undertaking this editing task was about pronouns: how should we refer to the Doctor in the wake of the Whittaker era? The question is not pedantry but is relevant to science, as Mike Stack demonstrates in his essay. Prior to Whittaker being cast as the Doctor in 2017, all *Doctor Who* texts referred to the Doctor using he/him/his pronouns, since the character had always been portrayed by male actors and was referred to on the show with male pronouns. But now, the long-awaited arrival of a female Doctor who is referred to on-screen with she/her/hers pronouns presents a grammatical challenge for scholars and commentators (discussed thoughtfully by McDunnah 2019). We, and the contributors, have been excited to address this, not least because some of us are trans, non-binary, genderqueer and/or genderfluid ourselves, as are some of our colleagues, comrades, friends and loved ones. Contributors have made their own decisions about specific instances of pronouns in their essays, sometimes using they/them/their throughout, and sometimes choosing she/her/hers and he/him/his when referring to specific incarnations of the Doctor, especially for the serials made between 1963 and 1989. But our editorial position was to insist on avoiding the old habit of using he/him/his as a generic, instead opting for they/them/their. This has at times created grammatically complex sentences, but in this we are reminded of an interchange from "The Two Doctors" (1985) between Colin Baker's Doctor and companions Peri and Jamie, regarding Troughton's Doctor:

> JAMIE: Now, my Doctor wouldnay' have done that.
> THE DOCTOR: Your Doctor is an antediluvian fogey. Allowing himself to be captured by the Sontarans! If anything happens to myself as a result of it, I will never forgive himself.
> PERI: Oh, I do wish you'd stop switching personal pronouns! It'd make it a lot easier to understand what you're talking about.

We consider it well within the spirit of *Doctor Who*—and arguably the optimistic spirit of science—to embrace and engage with the complexities of gender as a grammatical challenge, rather than considering the complexities a problem. Not limited to gender, *Doctor Who* has always been multi-dimensional in its outlook, even sometimes contradictory, because across the decades it has excitedly explored different nooks, crannies and alleyways of thought through many diverse perspectives and lenses. We hope you will find our book similarly diverse and complex in the perspectives it presents on *Doctor Who* and science.

REFERENCES

Bell, R. (2018, January 15) Event: Science of the Time Lords at National Space Centre. *Blogtor Who*. www.blogtorwho.com/event-science-time-lords-national-space-centre/ (accessed 9 December 2019).
Burk, G., and Smith, R. (2012) *Who Is the Doctor: The Unofficial Guide to Doctor Who: The New Series*. ECW/ORIM.
Carney, J. (2013) *Who's There: The Life and Career of William Hartnell*. Fantom.
Chapman, J. (2006) *Inside the TARDIS: The Worlds of Doctor Who: A Cultural History*. London: I.B. Tauris.
Charles, A. (2013) Three characters in search of an archetype: Aspects of the trickster and

the *flâneur* in the characterizations of Sherlock Holmes, Gregory House and Doctor Who. *Journal of Popular Television* 1(1): 83–102. doi:10.1386/jptv.1.1.83_1.
Cooke, L. (2003) *British Television Drama: A History*. London: Bloomsbury.
Coppa, F. (2010) Girl genius: Nyssa of Traken. In: Thomas, L.M., and O'Shea, T. (eds.) *Chicks Dig Time Lords: A Celebration of Doctor Who by the Women Who Love It*. Des Moines: Mad Norweigan Press, pp. 62–67.
Cox, B. (2013) *The Science of Doctor Who*. Britain: BBC Two.
Cranny-Francis, A. (2009) Why the Cybermen stomp: Sound in the new *Doctor Who*. *Mosaic* 42(2): 119–134.
de Kauwe, V., and Orthia, L.A. (2018) Knowledge, power and the ethics illusion: Explaining diverse viewer interpretations of the politics in classic era *Doctor Who*. *Journal of Popular Television* 6(2): 151–165. doi:10.1386/jptv.6.2.151_1
Dubois F-R (2015) Sciences imaginaires et imaginaire de la science dans *Doctor Who* (2005–2014). *Arts et Savoirs* 5: online. doi:10.4000/aes.333
Fiske, J. (1984) Popularity and ideology: A structuralist reading of *Dr. Who*. In: Rowland, W.D., Jr and Watkins, B. (eds.) *Interpreting Television: Current Research Perspectives*. Beverly Hills: Sage Publications, pp. 165–198.
Guerrier, S., and Kukula, M. (2015) *The Scientific Secrets of Doctor Who*. London: BBC Digital.
Harmes, M. (2016) Education in the fourth dimension: Time travel and teachers in the TARDIS. In: Readman, M. (ed.) *Teaching and Learning on Screen: Mediated Pedagogies*. London: Palgrave Macmillan, pp. 135–150.
Haynes, R. (2003) From alchemy to artificial intelligence: Stereotypes of the scientist in Western literature. *Public Understanding of Science* 12: 243–253.
Hearn, M. (2013) *Doctor Who: The Vault*. London: BBC Books.
IMDb (2019) William Hartnell Biography: Personal Quotes. *Internet Movie Database*. www.imdb.com/name/nm0367156/bio?ref_=nm_dyk_qt_sm#quotes (accessed 18 December 2019).
Johnson, D. (2013) Mediating between the scientific and the spiritual in *Doctor Who*. In: Crome, A., and McGrath, J.F. (eds.) *Religion and Doctor Who: Time and Relative Dimensions in Faith*. London: Darton, Longman and Todd, pp. 161–173.
Jones, R.A. (1997) The boffin: A stereotype of scientists in post-war British films (1945–1970). *Public Understanding of Science* 6: 31–48.
Jowett, L. (2014) The girls who waited? Female companions and gender in *Doctor Who*. *Critical Studies in Television* 9(1): 77–94. doi:10.7227/CST.9.1.6
Larsen, K. (2013) "They hate each other's chromosomes": Eugenics and the shifting racial identity of the Daleks. In: Orthia, L. (ed.) *Doctor Who and Race*. Bristol: Intellect, pp. 233–250.
Larsen, K. (2017) Ape-man or regular guy? Depictions of Neanderthals and Neanderthal culture in *Doctor Who*. In: Fleiner, C., and October, D. (eds.) *Doctor Who and History*. Jefferson: McFarland, pp. 148–167.
Larsen, K. (2018) The river, the rock, the relative and the returned: Depictions of women scientists in *Doctor Who*'s Moffat era. In: Carlson, A.L. (ed.) *Women in STEM on Television*. Jefferson: McFarland, pp. 187–206.
Layton, D. (2014) *The Humanism of Doctor Who: A Critical Study in Science Fiction and Philosophy*. Jefferson: McFarland.
McDunnah, M.G. (2019, January 16) Doctor He, She, or They? Changing gender, and language, in *Doctor Who*. *Conscious Style Guide*. consciousstyleguide.com/doctor-he-she-or-they-changing-gender-and-language-in-doctor-who/ (accessed 24 October 2019).
Morgain, R. (2013) Mapping the boundaries of race in *The Hungry Earth/Cold Blood*. In: Orthia, L. (ed.) *Doctor Who and Race*. Bristol: Intellect, pp. 251–267.
Oakes, K. (2013, November 21) The real science of "Doctor Who." *BuzzFeed*. www.buzzfeed.com/kellyoakes/the-real-science-of-doctor-who (accessed 9 December 2019).
O'Brien, D. (2000) *SF/UK: How British Science Fiction Changed the World*. Reynolds and Hearn.
O'Connor, P. (2012) *The Science of Doctor Who*. USA: BBC America.
O'Keeffe, M. (2017) Riding the wave: Science fiction media fandom and informal science education. *Journal of Science Fiction* 1(3): 24–39.

Orthia, L.A. (2010) *Enlightenment was the Choice: Doctor Who and the Democratisation of Science*. Ph.D. Thesis, The Australian National University.

Orthia, L.A. (2011a) Antirationalist critique or fifth column of scientism? Challenges from Doctor Who to the mad scientist trope. *Public Understanding of Science* 20(4): 525–542. doi:10.1177/0963662509355899.

Orthia, L.A. (2011b) "Paradise is a little too green for me": Discourses of environmental disaster in *Doctor Who* 1963–2010. *Colloquy* 21: 56–80.

Orthia, L.A. (2013) Savages, science, stagism and the naturalized ascendancy of the Not-We in *Doctor Who*. In: Orthia, L. (ed.) *Doctor Who and Race*. Bristol: Intellect, pp. 269–287.

Orthia, L.A. (2019) How does science fiction television shape fans' relationships to science? Results from a survey of 575 *Doctor Who* viewers. *Journal of Science Communication* 18(04): A08. doi:10.22323/2.18040208.

Orthia, L.A., and Morgain, R. (2016) The gendered culture of scientific competence: A study of scientist characters in *Doctor Who* 1963–2013. *Sex Roles* 75: 79–94. doi:10.1007/s11199-016-0597-y.

Parsons, P. (2006) *The Science of Doctor Who*. Cambridge: Icon Books.

Patrick, R. (2017, August 22) I. can't explain how excited, I. was when Doctor Who got a black companion. *The Guardian*. www.theguardian.com/commentisfree/2017/aug/22/i-cant-explain-how-excited-i-was-when-doctor-who-got-a-black-companion (accessed 9 December 2019).

Powers, T. (2016) *Gender and the Quest in British Science Fiction Television*. Jefferson: McFarland.

Richmond, A. (2015, September 18) Doctor Who vs real world science: Who comes up trumps? *The Conversation*. theconversation.com/doctor-who-vs-real-world-science-who-comes-up-trumps-45183 (accessed 9 December 2019).

Robb, B.J. (2009) *Timeless Adventures: How Doctor Who Conquered TV*. Harpendin: Oldcastle Books.

Tulloch, J., and Alvarado, M. (1983) *Doctor Who: The Unfolding Text*. London: Macmillan Press.

Tulloch, J., and Jenkins, H. (1995) *Science Fiction Audiences: Watching Doctor Who and Star Trek*. London: Routledge.

Vohlidka, J. (2013) Doctor Who and the critique of western imperialism. In: Orthia, L. (ed.) *Doctor Who and Race*. Bristol: Intellect, pp. 123–139.

White, M. (2005) *A Teaspoon and an Open Mind: The Science of Doctor Who*. London: Allen Lane.

Williams, R. (2014, March 22) The Science of Doctor Who. *The Science Show*. www.abc.net.au/radionational/programs/scienceshow/the-science—of-doctor-who/5337704 (accessed 9 December 2019).

Who's Planet Looks Like Home?

J.J. ELDRIDGE

One of the amazing things about early science fiction is the widespread assumption that there were planets around other stars. Many of the early TV sci-fi series had people traveling between planets, decades before the first planet that orbited another typical star like our Sun was confirmed by Mayor and Queloz (1995), soon after the classic *Doctor Who* series came to the end of its run in 1989.

Of course, there had been much speculation, with the earliest recorded being that of Giordano Bruno in 1584 (Maor 1987) and later Isaac Newton in 1713 (Newton et al. 1999). These natural philosophers took the first difficult step of realizing that stars are other Suns and thus likely to hold dominion over similar planetary systems. More recently, Struve (1952) suggested that many of the other planets might be very different to our own.

Nearly all early TV science fiction series were set predominantly either on Earth or Earth-like habitable planets.[1] There are good reasons for this; first, to attract a broad range of viewers it was thought that familiar locations were required. Second, it was easier to visit a planet that looked like Earth rather than attempt to create a very alien world, although examples of where this was done do occur within *Doctor Who*, for example "The Web Planet" (1965).

In this essay, I explore whether our developing understanding of the Universe over the significant broadcast run of the classic series was reflected on screen, and if the discovery of planets around other stars caused any change in the planets depicted in the new series of *Doctor Who*.

I first consider our Solar System and how early incorrect ideas on how the planets formed can explain why it was thought Mars could host complex life like the Ice Warriors. I then move on to consider exoplanets (planets around other stars), a field of study that has shown a dramatic growth in knowledge between the two series of *Doctor Who*. When the classic series of *Doctor Who* ended there were suggestions that there were planets around other stars (Campbell et al. 1988), but today we know in excess of 4104 (NASA exoplanet archive 2019). I also explore how today's researchers are grappling with concepts around colonization and how we might deal with the competing demands of wanting new worlds to inhabit, set against industrial concerns and the prior claim of any indigenous life we might find.

I also consider some of the weirder planets of *Doctor Who* and show how our current understanding of astrophysics suggests some of these fictional planets are more likely to exist in the Universe than one might naively expect. I then present evidence that, while not widespread, *Doctor Who* has certainly inspired a few astrophysicists in their research.

Earth and Similar Planets in Doctor Who

Earth and Earth-like planets are common in *Doctor Who*. To estimate the percentage of visits to different locations in the series I considered each televised episode (rather than each story or serial) of *Doctor Who* and judged the nature of that planet from the on-screen appearance, sometimes extrapolating based on comments in the episodes. In the graph, I show the percentage of episodes in different locations. My classifications were:

1. Earth.
2. Earth-like habitable planets with life and a breathable atmosphere.
3. Earth-like planets—apparently Earth-like, Earth-gravity, a nearby star, a breathable atmosphere but no signs of indigenous life.
4. Moons or uninhabitable planets—with no or minimal atmosphere.
5. Space—i.e. set in space stations or spacecraft.
6. Nowhere—the strange episodes that do not occur at any specific location, for example "The Celestial Toymaker" (1966) or "Amy's Choice" (2010).
7. Weird—a planet that is quite fantastic and requires novel physics to explain.

The relative numbers in Figure 1 show that overall the Doctor has spent more than 50 percent of their time visiting Earth (mainly within the United Kingdom), with another 20 percent of their time visiting Earth-like habitable worlds. The remaining time is in space, visiting less habitable locations and more rarely other settings. There is also a notable difference between the classic and new era of *Doctor Who*, with Earth visited more often in the new series than the old. In the new series, space is visited more commonly

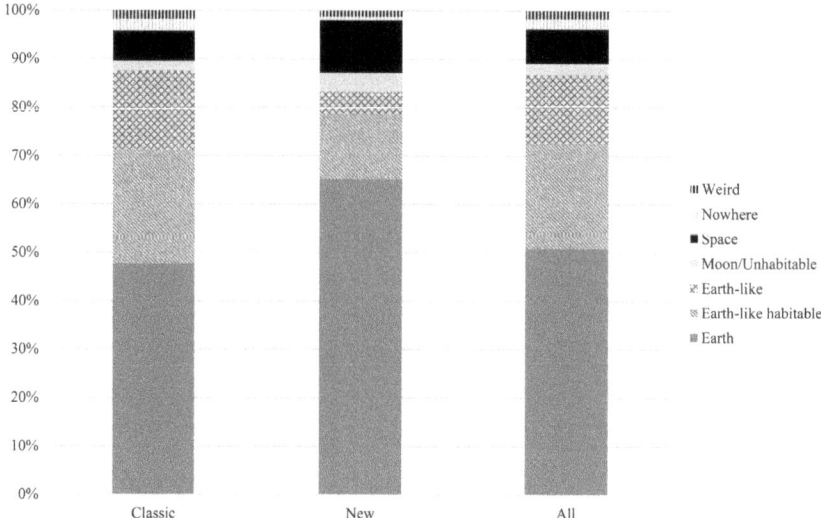

Figure 1. The relative percentage of episodes in different locations through the Universe in the classic series, from "An Unearthly Child" (1963) to the "Doctor Who Movie" (1996), in the new series, from "Rose" (2005) to "Resolution" (2019), and all episodes from both samples combined.

than other Earth-like worlds, while the reverse is true for the classic series. I explore the reasons for this later.

The Solar System

With the depth of our knowledge in the twenty-first century it is easy to forget that just one century ago we didn't know very much about the Universe. Today we know the Universe is 13.8 billion years old (Planck Collaboration 2018), the Sun is 4.6 billion years old (Bonanno et al. 2002), while the Earth is only 4.54 billion years old (Dalrymple 2001). The precision of these numbers shows we are living in the age of precision astrophysics and cosmology. We can observe the distant Universe, seeing back in time to within a few hundred million years after the Big Bang. But what did, or rather did not, we know about the Universe back in November 1963? Also, how much scientific knowledge was familiar to the general public, especially the writers of *Doctor Who*? It is difficult to be certain, looking back, but we can separate out what we knew about the stars, galaxies and the Universe from what we knew about our Sun and the Solar System.

We knew that stars lived for a very long time: Eddington (1920) had suggested that nuclear fusion at the center of stars would allow them to shine for billions of years. We also knew the structure of our own Galaxy and that our Sun was just one of hundreds of billions. We had begun to understand that the composition of the Sun and stars was very different to planets: Payne (1925) had discovered that the Sun and stars are mostly made of hydrogen, helium and very little of the other elements.

We also knew due to the work of Edwin Hubble that our Galaxy was one of many; that the Universe was expanding and had formed in a "Big Bang" (Hubble 1929). However this was only firmly established in 1964, after *Doctor Who*'s first season, when Penzias and Wilson (1965) observed the cosmic microwave background radiation, the left-over cooling light from the Big Bang.

By contrast, in 1963 we were not certain about how our Solar System had formed. The first realistic model was put forward by Rene Descartes in 1632 (Williams and Cremin 1968). But the correct model was first put forward by Emanuel Swedenborg (1734) and was later built upon by Immanuel Kant in 1755 and then Pierre-Simon Laplace in 1796 (See 1909). The key idea postulated by these authors was that initially a giant nebula of gas and dust collapsed under its own gravity, spinning up to form a disk that then became our Solar System. However, if this was the case this suggests that the Sun should be rapidly rotating. The Sun rotates approximately once a month[2] but this is slower than the simple nebula model would predict and so it must have slowed down.

The biggest leaps in our understanding of the Solar System were taken during the time *Doctor Who* was airing. Safronov (1972) and Prentice (1978) revived the Laplacian nebular model, updating it and including physics describing how the disk from which the planets form will slow down the Sun. Their work became the Solar Nebular Disk Model. Even though there are outstanding problems, this is at the core of the best model today.

It's significant that as *Doctor Who* aired our understanding increased, but we see no change in how the Solar System is portrayed within the series. In the classic series the Doctor was very much visiting either Earth, its Moon (see Stanway, this volume) or some other Earth-like world. The two exceptions are in "The Invisible Enemy" (1977) when people visited an interplanetary filling station on Saturn's moon Titan. Then in the same year the Doctor visited Pluto, which resembled Earth, in "The Sun Makers" (1977) but little was discussed about general properties of the Solar System.

An exception and a planet worthy of special consideration here is Mars. The idea of intelligent life on Mars was kindled after the observations of Giovanni Schiaparelli's "canali" were built upon to become Percival Lowell's canals (Lowell 1906). A mistranslation and an optimistic extrapolation led Lowell to write, "That Mars is inhabited by beings of some sort or other we may consider as certain as it is uncertain what those beings may be" (376). Resulting speculation in science fiction was rife, most notably appearing in *The War of the Worlds* (1897) by H.G. Wells. *Doctor Who* used the idea to create the Ice Warriors: Martians who were ready to invade Earth to escape their dying cold world in "The Ice Warriors" (1967) and "The Seeds of Death" (1969).

While *Doctor Who* was going to air, the promise of life on Mars had become far less certain than Lowell believed. However, the writers would certainly have grown up with the idea being discussed and popularized, thus it is no surprise that life on Mars was a key assumption of the *Doctor Who* canon.

It is significant that given this early understanding of the Solar System the idea of a technologically superior race on Mars was not so strange. One early model of the planets' formation was that to slow down the Sun, planets fissioned or were "thrown" out from the Sun. The next logical inference was that planets further away from the Sun formed earlier than those closer to the Sun and thus had more time for life to evolve (e.g., Jacot 1986).

Today we know that Mars is very similar in age to the Earth. We also know the Lowell's canals do not exist (Sagan and Fox 1975); but we also know that in the ancient past, billions of years ago, it was habitable and had vast oceans, seas and lakes (e.g., McLennan and Grotzinger 2008; Ehlmann et al. 2011; Matsubara et al. 2011; Onstott et al. 2019). However, this period was very short because Mars is much smaller than Earth. It is a tenth the mass and half the radius. This means its interior cooled more quickly than Earth's and the core's magnetic dynamo shut down, no longer producing a protective planetary magnetic field. Today Mars is a cold, dead world with an atmosphere only a hundredth the density of Earth's, as the absence of a planetary magnetic field means the atmosphere is constantly ablated by the solar wind.

This knowledge is relatively recent, from the flotilla of missions to Mars, which reveal its ancient past. Thus, while there is clearly not complex life on Mars today, there may well have been simple life in the past, and there may still be simple life there today. Therefore, we continue to return to this planet. It also indicates that maybe there is some basis in fact for the Ice Warriors. If life is quick to develop and evolve into humanoid life forms then Martians may have existed in the past, although again there is no evidence for that in reality. The notion of ancient life on Mars does however resonate with some aspects of the Martian culture shown in *Doctor Who*. While the Martians and their lords are aggressors in "The Ice Warriors" and "The Seeds of Death," in "The Curse of Peladon" (1972) they are reconfigured as a more noble and distinguished civilization.

It is worth noting here that in the *Doctor Who* universe it does appear life is quite quick to evolve to complex civilizations. On Earth humans are the second complex civilization to appear with the Silurians and Sea Devils being the first as described in the novelization of "The Silurians" (1970): *Doctor Who and the Cave Monsters* by Hulke (1974). The Sea Devils are revealed as amphibious cousins to the Silurians in "The Sea Devils" (1972). Again, there is no evidence this is the case in reality, but it is an interesting thought experiment to explore what our Solar System could be like if life was that little bit faster in evolving.

Furthermore, the fact that we have found water on Mars has fed back into *Doctor Who*, leading directly to the quite scary "The Waters of Mars" (2009). This story perhaps better represents the future history of Mars when humanity will one day land on Earth's neighboring world, subsist on its soils or even terraform it, to make it truly Earth-like.

Planets Around Other Stars

The early history of detection of planets around other stars is quite contentious because it is difficult to do! Planets do not shine themselves and we can only detect them via the reflected light of their host star or by some other indirect method. The first suggestion that there were planets around other stars, referred to by astrophysicists as an exoplanet, was by Campbell, Walker and Yang (1988) and these early candidates have now been confirmed (Hatzes et al. 2003). The first confirmed planet was from Wolszczazn and Frail (1992), however this planet orbits a dead, neutron star, the remnant left over from when a star dies in a supernova.

The first widely accepted planet discovery around a long-lived star like our Sun (where we might expect life to be possible) was 51 Pegasi b, discovered by Mayor and Queloz (1995). It is a strange world though: it has the mass of Jupiter but in a 4.2-day orbit, placing it close to the star. It was this discovery that indicated planets could be around every star and we began looking in earnest.

The difficulty in discovering planets is that they are hundreds of times smaller in radius and hundreds of thousands of times less massive than stars. The first planets were identified by radial velocity variations. This is where, as a planet orbits around its host star, the star also orbits around a much smaller circle and the light of star's spectrum slowly blueshifts and redshifts as its motion varies towards and away from us. This is a difficult method as the wobble of a star's orbit is in the order of meters per second and so requires the most extreme precision to detect. Today most planets are instead found by detecting the slight decrease in luminosity of a star as they pass in front. Again, as most planets are 100 times smaller than their host star, this requires measuring how the light changes, to 0.01 percent or better! After nearly three decades of practice we've become pretty good at it. We are now at a stage where we have been able to identify many types of planets that we don't have in our own Solar Systems such as super-Earths, hot Jupiters and mini-Neptunes.

Planet finding has now become an industrial-scale endeavor and it is difficult to keep up with the exact known number of exoplanets, with the number at writing now 4104 (NASA Exoplanet archive 2019). Many of these are the only known planet around a star but there are 669 known multi-planet systems. The number of detected exoplanets has increased as we have built better techniques and machines for discovering planets. Most of these systems have been discovered over the same time span as the new series of *Doctor Who*.

The most interesting discovery perhaps, especially from the viewpoint of science fiction fans, is the discovery of planets orbiting binary star systems. *Doctor Who* was one of the earliest to depict on screen binary star systems, both in "The Chase" (1965) and "The Krotons" (1968–69). The Doctor also explicitly stated that Gallifrey is a super-Earth planet orbiting a binary star system in "Gridlock" (2007), which was broadcast only a few years after the first discovery of such a planet (Correia et al. 2005).

One might have expected therefore such planets to pick up the nickname "Gallifrey" but this is not the case. The early black and white episode depictions of binary suns were quick, simple and will forever be overshadowed by the haunting beauty of the binary sunset on Tatooine from *Star Wars: A New Hope* (1977), especially with the accompanying music score, leading to such systems being referred to colloquially as "Tatooines."

Given the rich variety of new planetary systems we have learnt of in our Galaxy it becomes a surprise to realize that in the new series of *Doctor Who*, as shown in Figure 1, more time is spent visiting Earth than any other worlds.

A hint for the reasoning behind this circumstance comes from thinking about the classic series, and what happened during the 1970 and 1986 seasons. In the 1970 season, the Time Lords stranded the Doctor on Earth as punishment. This was a move designed to increase the number of viewers and it did so with some success if we consider viewer figures (*Doctor Who News* 2019). However, the reverse happened in the season-long serial from 1986, "The Trial of a Time Lord": Earth was not featured at all (although it was revealed that the planet Ravolox was Earth in the far future, ravaged by solar flares, and on screen the characters' knowledge of their planet's history had been suppressed by the Time Lords). This season saw a significant drop in the number of viewers. While there are many factors suggested for this drop, it is likely that many viewers didn't enjoy a series that didn't appear to relate to the world around them that lacked frames of reference anchored in daily reality.

This could explain why the new series of *Doctor Who* has so many Earth-based episodes, to keep it grounded with viewers' everyday experience. Looking more closely at the data it is possible to see that the new series is now starting to increasingly visit other planets, but I expect the number will never exceed 50 percent. This is a shame given that today we know of increasing numbers of Earth-like worlds and it is only a matter of time before we begin to learn more about their atmospheres and suitability for life. However perhaps studying planets in the Galaxy and seeing that Earth-twins are rare makes us realize that our own planet is far more distinctive than we might expect. It is worth exploring and defending, and thus worthy of many a story set in its beautiful landscapes.

Colonization and Terraforming

An important aspect of representations of planets in society and in science fiction concerns who owns a world. Who has the right to inhabit one and colonize it? Western civilization has a very bad track record of how it has treated indigenous peoples when expanding its territory on Earth. When we look at the Universe, we have so far seen it barren and lifeless. We might at first think it is all there for the taking as *terra nullius*? However, is that an ethical viewpoint if we think more deeply? Sachs (2018) summarized the debate around exploration and inhabiting the Universe, while Levchenko and colleagues (2018) discuss the specific questions we face for going to Mars today in detail.

Even if a planet is lifeless, whether it is worthwhile to go there or not is an interesting question. We could theoretically go to Mars and terraform it into a more pleasant environment for humans to live in. However, any people we send face severe risks to their physical and mental health. Can we ask people, even if they are fully informed, to take such risks?

Then if we do discover life on Mars, even if it is just microbes, what do we do? Do we ignore Mars and leave it for the microbes, with the assumption that in millions of years as the Sun evolves it will once more become suitable for complex life to evolve? Or do we change the habitat to one suitable for our own existence? Such a question will likely be asked of us as some point in the future and *Doctor Who* has yet to consider it. The situation is taken quite seriously by space scientists today with planetary protection, of other planets, leading to strict guidelines for missions involving planetary landings (e.g., NASA 2019).

It is worth pointing out that humanity has not been so keen on habitat modification when encountering it in science fiction: in H.G. Wells' *War of the Worlds* and in the *Doctor Who* serial "The Seeds of Death," the Martians and Ice Warriors are trying to Areform (recreate a Martian environment on) Earth to be more habitable for them. The Sontarans do the same in "The Sontaran Stratagem/The Poison Sky" (2008). The plot of these stories indicates how unpleasant it is to be on the other end of a redevelopment of your planet's climate and ecosystem.

In "The Mutants" (1972) the Doctor shows their opposition to such geocidal terraforming by humans. In the new series there have been a few episodes mentioning terraforming of planets to make them suitable for human habitation. However, none have really gone into detail compared to *Star Trek: The Next Generation* (1987–94) where an entire episode, "Home Soil" (1988), was given over to this very question. In that case, the terraforming of a planet was halted, and the simple crystalline life forms allowed to continue their existence without inference.

The classic series of *Doctor Who* did examine a much more tangible dilemma of a potentially habitable and terraformable planet with extant indigenous inhabitants and a complication from industrial concerns. In "Colony in Space" (1971) the Third Doctor visits a world where colonists are struggling to make their settlement self-sustaining, the indigenous humanoid population have devolved owing to the malign influence of a doomsday weapon that has blighted the planet, while an unscrupulous mining company ship is purely after the mineral wealth of the planet. In theory, all three could live peaceably together, but the mining company's greed soon throws many spanners in the works. While the focus is on capitalism rather than the ethics of settlement, this story highlights the very real issues that we need to consider if humanity ever move beyond the cradle of our own planet.

Already today, such issues are being discussed within space exploration (Sachs 2018; Levchenko et al. 2018). I consider that the episodes of *Doctor Who* discussing such issues should be treated as thought experiments, usually of a worst-case scenario that can teach us why we need to consider the ethics of space exploration and get it right. Orthia's recent survey (2019) suggests that such episodes of *Doctor Who* do make viewers think more deeply the ethics of science and exploration as well as the need to protect our habitable environment on Earth itself.

Ethics is certainly something *Doctor Who* has returned to frequently over its run. It is interesting, given all the issues highlighted with colonizing and visiting other planets discussed by Levchenko and colleagues (2018), that *Doctor Who* has always had the continuing theme that it is the fate of humans to explore and spread across the Galaxy. Hopefully we'll take care of those new worlds we visit.

Weird Planets

A fun aspect of *Doctor Who* is the sometimes very *weird* planets visited. Here our definition of a weird planet is one quite unlike those in our Solar System or outside the common population of planets we know of scientifically: either in location, composition or some other characteristic.

The three weirdest planets, in my opinion, are Zanak from "The Pirate Planet" (1978), Voga from "Revenge of the Cybermen" (1975) and Krop Tor from "The Impossible Planet/The Satan Pit" (2006). How viable and realistic are each of these? What does our latest understanding of astrophysics tell us about their nature?

Zanak is a planet that the Captain has turned into a spaceship. To construct such a planet is a vast feat of extreme engineering (described on screen as an "exquisite exercise in gravitational geometry"), and that is even before we consider how you'd make it travel through space! Zanak was equipped with a warp engine so would appear around another planet, thus the problem of how to stay warm traveling between stars didn't have to be solved. This is something the Daleks would have had to do if they had succeeded in "The Dalek Invasion of Earth" (1964) to turn the Earth into a similar spaceship. As extreme as these ideas sound, it has been suggested that we should check modern astrophysics surveys of the sky for such planetary spaceships (Forgan 2013).

By comparison, Voga, the gold planet, which is key to the defeat of the Cybermen in a war after the humans had invented the glitter gun, is more reasonable. Gold is the one weakness of the Cybermen's respiratory system as has been repeatedly demonstrated within the series. But how likely is a planet entirely made of gold? This links back to our discussion above, concerning how solar systems form. Most planets form out of a diffuse nebula which is a mix of materials, although still predominately hydrogen and helium; a completely gold planet is unlikely to form in typical solar systems like ours.

However, the first confirmed planets were discovered in a very different location, in orbit around a neutron star, the stellar remnant left over after a massive star has exploded in a supernova (Wolszczan and Frail 1992). A neutron star can have its motion tracked extremely accurately and this allowed astronomers to deduce that a system of planets was in orbit around neutron star PSR1257 +12. The planets must have formed *in situ* as any present before the explosion would have been destroyed or flung away from the star during the supernova. These planets would thus be enriched in the materials that supernovae produce such as oxygen, silicon and iron-group elements, but still very little gold. However, if planets can form after an explosive supernova maybe they can form after other types of stellar explosions?

In 2017, the merger of two neutron stars was detected by gravitational waves and light from across the electromagnetic spectrum (LIGO, Virgo and other collaborations 2017). This likely formed a black hole but also threw out a lot of neutron star material (or neutronium), which formed from many heavy elements. Possibly up to a Jupiter mass worth of gold was formed, along with other similar elements such as silver and platinum. A world such as Voga might result from such an event and would orbit the newly formed black hole born in the catastrophe.

Which brings us to the final Weird planet, Krop Tor. This planet is in orbit around a black hole. While on screen it's described as being in an impossible orbit, it is not clear if this is true as it does appear to be well outside the radius of the innermost stable orbit permitted by general relativity (e.g., Misner, Thorne and Wheeler 2017). As mentioned above we know of planets orbiting neutron stars so it is reasonable to expect that planets will also orbit black holes as they are formed in similar supernovae. Perhaps if the black hole was formed by a neutron star merger rather than a supernova, it might even be a gold planet. Will this always be hypothetical? Or is there ever a chance that we will discover a planet orbiting a black hole? Detections are difficult simply because a black hole emits no light. Current planet-finding methods mostly rely on the light of the host star rather than the planet itself. Thus, on first thought the answer might be no.

But with the power of *Doctor Who* to inspire, can the premise from *Doctor Who* that a black hole might have an orbiting planet, linked to circumstantial evidence from known neutron star planets, inspire us to suggest a way to detect planets around black holes?

The answer is yes! We can use a well-established planet-finding technique called gravitational microlensing (Beaulieu et al. 2006). If an object is massive enough its gravity can cause light to follow a curved path. Our own Sun does this, and the star field observed around the Sun during an eclipse would look slightly warped compared to the same patch of sky 6 months later without the Sun in the way.

If a large object somewhere in the Galaxy passes between our line of sight and a background star, the object's gravitational field will lens the background star's light and will make it look brighter. If the object in the middle is a star with planets these can also cause lensing, although it will be weaker than that of the star. We have found many planets this way, including free-floating planets without a host (Mróz et al. 2017).

In searches for planets using this technique, astronomers have already found many candidate black holes, so we could begin searching for planets around these objects (Wyrzykowski et al. 2016). The issue however is a timescale problem. A stellar lensing event has a timescale of a few months, while a black hole has a timescale of several months and planets of only days at most (Maoz 2007). We will of course one day get lucky, but just this thought is enough to start people sifting through the data we are now obtaining in ever greater quantities.

Impact of Doctor Who on Astronomy and Astrophysics

For most of this essay, we have discussed the depiction of planets within *Doctor Who* and highlighted how this has changed over time along with our understanding of our own Solar System and planets around other stars. For the most part, even with the excess of habitable planets, the two seem to correspond, although there is always a strong bias towards visiting Earth to make the stories relevant for viewers.

It is interesting however to ask the reverse question: has the depiction of planets in *Doctor Who* inspired astronomers or astrophysicists in their work? The answer is yes, although not directly on planets. Even the outline presented above indicating how planets around black holes could be detected is a clear case of *Doctor Who* inspiring a quick bit of original science. Orthia (2019) performed a survey of science-interested *Doctor Who* viewers and found that watching *Doctor Who* possibly affected 25 percent (those that answered yes or maybe) of the respondents' education choices. As many as 20 percent believed that their long-term career choices were affected. This implies a significant impact on society, encouraging more people into STEM fields.

Anecdotal evidence also suggests that a large fraction of astronomers have been engaged and enthused by science fiction, and in the United Kingdom by *Doctor Who*. This has certainly been a motivating factor for both astronomers writing in this book, Elizabeth Stanway and the current author, both of whom have also used science fiction scenarios for teaching and science communication. This interest also partly inspired the work of Stanway on the variation of habitability with cosmic time. Stanway and colleagues (2018) investigated how the habitability of whole galaxies has evolved with time, showing that the Universe only really started to become habitable for life around 5 billion years ago, close to the birth epoch of our own planet and Sun. While there is nothing in the paper that specifies *Doctor Who*, as an inspiration the author has also contributed an essay to this volume and is inspired by her interest in sci-fi, especially *Doctor Who*.

More direct evidence of the influence of *Doctor Who* on the field is seen in "the TARDIS papers," two articles that were published online at the ArXiv.org depositary for scientific texts. They include a detailed scientific paper where the authors describe a "Traversable Achronal Retrograde Domain in Spacetime" that would allow time travel both into the past and the future (Tippett and Tsang 2013a), and a more accessible lay summary (Tippett and Tsang 2013b). Here the figures are explicitly inspired by *Doctor Who* and illustrate a rigorous, if frivolous, attempt to put forward solutions of General Relativity that allow for time travel.

A final example is the use of "TARDIS" as the name for a computer program that models the spectra from supernovae (Kerzendorf and Sim 2014). The authors chose this name purely because they liked *Doctor Who*. These few examples indicate that *Doctor Who* has inspired many people to pursue a scientific career and ask difficult questions. I certainly know this is at least true for me and I made sure to thank *Doctor Who* in the acknowledgments of my Ph.D. thesis (Eldridge 2005)!

Notes

1. In this essay I use the term "Earth-like" planets to refer to those of similar size and orbit to Earth. Those with extant life on the surface and a breathable atmosphere are referred to as "Earth-like habitable" planets.
2. As the Sun is not solid its rotation is complex. We know the equator rotates once every 25 days while the poles rotate once every 33 days.

References

Beaulieu, J.P., Bennett, D.P., Fouqué, P., Williams, A., Dominik, M., et al. (2016) Discovery of a cool planet of 5.5 Earth masses through gravitational microlensing. *Nature* 439: 437–440. doi:10.1038/nature04441.
Bonanno, A., Schlattl, H., and Paternò, L. (2002) The age of the Sun and the relativistic corrections in the EOS. *Astronomy and Astrophysics* 390: 1115–1118. doi:10.1051/0004-6361:20020749.
Campbell, B., Walker, G.A.H., and Yang, S. (1988) A search for substellar companions to solar-type stars. *The Astrophysical Journal* 331: 902–921. doi:10.1086/166608.
Correia, A.C.M, Udry, S., Mayor, M., Laskar, J., Naef, D., Pepe, F., Queloz, D., and Santos, N.C. (2005) The CORALIE survey for southern extra-solar planets. XIII. A pair of planets around HD 202206 or a circumbinary planet? *Astronomy and Astrophysics* 440: 751–758. doi:10.1051/0004-6361:20042376.
Dalrymple, G.B. (2001) The age of the Earth in the twentieth century: A problem (mostly) solved. *Special Publications, Geological Society of London* 190: 205–221. doi:10.1144/GSL.SP.2001.190.01.14.
Doctor Who News (2019) *Doctor Who Guide*. guide.doctorwhonews.net/ (accessed 7 December 2019).
Eddington, A.S. (1920) The Internal Constitution of the Stars. *The Scientific Monthly* 11: 297–303.
Ehlmann, B.L., Mustard, J.F., Murchie, S.L., Bibring J-P, Meunier, A., Fraeman, A.A., and Langevin, Y. (2011) Subsurface water and clay mineral formation during the early history of Mars. *Nature* 479: 53–60. doi:10.1038/nature10582.

Eldridge, J.J. (2005) *Progenitors of Core-Collapse Supernovae*. Ph.D. Thesis, University of Cambridge. https://arxiv.org/pdf/astro-ph/0502046.pdf.
Forgan, D.H. (2013) On the Possibility of Detecting Class A Stellar Engines using Exoplanet Transit Curves. *Journal of the British Interplanetary Society* 66: 144–154.
Hatzes, A.P., Cochran, W.D., Endl, M., McArthur, B., Paulson, D.B., Walker, G.A.H, Campbell, B., and Yang, S. (2003) A Planetary Companion to Gamma Cephei A. *Astrophysical Journal* 599: 1383–1394. doi:10.1086/379281.
Hubble, E. (1929) A relation between distance and radial velocity among extra-galactic nebulae. PNAS 15: 168–173. doi:10.1073/pnas.15.3.168.
Hulke, M. (1974) *Doctor Who and the Cave Monsters*. BBC Books.
Jacot, L. (1986). *Heretical Cosmology* (transl. of Science et bon sense, 1981). Exposition-Banner.
Kerzendorf, W.E., and Sim, S.A. (2014) A spectral synthesis code for rapid modelling of supernovae. *Monthly Notices of the Royal Astronomical Society* 440: 387–404. doi: 10.1093/mnras/stu055.
Levchenko, I., Xu, S., Mazouffre, S., Keidar, M., and Bazaka, K. (2018) Mars Colonization: Beyond Getting There. *Global Challenges* 3: 1800062. doi:10.1002/gch2.201800062.
LIGO, Virgo and other collaborations (2017) Multi-messenger Observations of a Binary Neutron Star Merger. *The Astrophysical Journal* 848: L12. doi:10.3847/2041-8213/aa91c9.
Lowell, P. (1906) *Mars and Its Canals*. The Macmillian Company, London.
Maor, E. (1987) *To Infinity and Beyond: A Cultural History of the Infinite*. Boston: Birkhäuser.
Maoz, D. (2007) *Astrophysics in a Nutshell*. Princeton University Press.
Matsubara, Y., Howard, A.D., and Drummond, S.A. (2011) Hydrology of early Mars: Lake basins. *Journal of Geophysical Research* 116(E04): 001. doi:10.1029/2010JE003739.
Mayor, M., and Queloz, D. (1995) A Jupiter-mass companion to a solar-type star. *Nature* 378: 355–359. doi:10.1038/378355a0.
McLennan, S.M., and Grotzinger, J.P. (2008) The sedimentary rock cycle of Mars. In: Bell, J. (ed.) *The Martian Surface—Composition, Mineralogy, and Physical Properties*. Cambridge University Press, pp. 541–577.
Misner, C.W., Thorne, K.S., and Wheeler, J.A. (2017) *Gravitation*. Princeton University Press.
Mróz, P., Udalski, A., Skowron, J., Poleski, R., Kozłowski, S., Szymański, M.K., Soszyński, I., Wyrzykowski, Ł., Pietrukowicz, P., Ulaczyk, K., Skowron, D., and Pawlak, M. (2017) No large population of unbound or wide-orbit Jupiter-mass planets. *Nature* 473: 349–352. doi:10.1038/nature23276.
NASA (2019) Planetary Protection Board report. www.nasa.gov/sites/default/files/atoms/files/planetary_protection_board_report_20191018.pdf (accessed 7 December 2019).
NASA Exoplanet Archive (2019) exoplanetarchive.ipac.caltech.edu/ (accessed 7 December 2019).
Newton, I., Cohen, B., and Whitman, A. (1999) *The Principia: A New Translation and Guide*. University of California Press.
Onstott, T.C., Ehlmann, B.L., Sapers, H., Coleman, M., Ivarsson, M., Marlow, J.J., Neubeck, A., and Niles, P. (2019) Paleo-Rock-Hosted Life on Earth and the Search on Mars: A Review and Strategy for Exploration. *Astrobiology* 19: 1230–1262. doi:10.1089/ast.2018.1960.
Orthia, L.A. (2019) How does science fiction television shape fans' relationships to science? Results from a survey of 575 *Doctor Who* viewers. *Journal of Science Communication* 18(04): A08. doi:10.22323/2.18040208.
Payne, C.H. (1925) *Stellar Atmospheres; a Contribution to the Observational Study of High Temperature in the Reversing Layers of Stars*. Ph.D. Thesis, Radcliffe College.
Penzias, A.A., and Wilson, R.W. (1965) A Measurement Of Excess Antenna Temperature At 4080 Mc/s. *Astrophysical Journal Letters* 142: 419–421. doi:10.1086/148307.
Planck Collaboration (2018). Planck 2018 results. VI. Cosmological parameters. *arXiv*:1807.06209. arxiv.org/abs/1807.06209 (accessed 7 December 2019).
Prentice, A.J.R. (1978) Origin of the solar system. I—Gravitational contraction of the turbulent protosun and the shedding of a concentric system of gaseous Laplacian rings. *The Moon and the Planets* 19: 341–398. doi:10.1007/BF00898829.
Sachs, B. (2018) Eight ethical questions about exploring outer space that need answers. *The Conversation*. theconversation.com/eight-ethical-questions-about-exploring-outer-space-that-need-answers-98878 (accessed 11 December 2019).

Safronov, V.S. (1972) *Evolution of the Protoplanetary Cloud and Formation of the Earth and the Planets*. Israel Program for Scientific Translations.

Sagan, C., and Fox, P. (1975) The Canals of Mars: An Assessment After Mariner 9. *Icarus* 25: 602–612. doi:10.1016/0019-1035(75)90042-1.

See, T.J.J. (1909) The Past History of the Earth as Inferred from the Mode of Formation of the Solar System. *Proceedings of the American Philosophical Society* 48: 119–128.

Stanway, E.R., Hoskin, M.J., Lane, M.A., Brown, G.C., Childs, H.J.T, Greis, S.M.L, Levan, A.J. (2018) Exploring the cosmic evolution of habitability with galaxy merger trees. *Monthly Notices of the Royal Astronomical Society* 475: 1829–1842. doi:10.1093/mnras/stx3305.

Struve, O. (1952) Proposal for a project of high-precision stellar radial velocity work. *The Observatory* 72: 199–200.

Swedenborg, E. (1734) *Opera Philosophica et Mineralia. Principia* 1.

Tippet, B.K., Tsang, D. (2013a) Traversable Achronal Retrograde Domains In Spacetime. arXiv:1310.7985 arxiv.org/abs/1310.7985 (accessed 11 December 2019).

Tippet, B.K., Tsang, D. (2013b) The Blue Box White Paper. arXiv:1310.7983 arxiv.org/abs/1310.7983 (accessed 11 December 2019).

Williams, I.O., and Cremin, A.W. (1968) A survey of theories relating to the origin of the solar system. *Quarterly Journal of the Royal Astronomical Society* 9: 40–62.

Wolszcazn, A., and Frail, D. (1992) A planetary system around the millisecond pulsar PSR1257 + 12. *Nature* 355: 145–147. doi:10.1038/355145a0.

Wyrzykowski, Ł., Kostrzewa-Rutkowska, Z., Skowron, J., Rybicki, K.A., Mróz, P., Kozłowski, S., Udalski, A., Szymański, M.K., Pietrzyński, G., Soszyński, I., Ulaczyk, K., Pietrukowicz, P., Poleski, R., Pawlak, M., Iłkiewicz, K., and Rattenbury, N.J. (2016) Black hole, neutron star and white dwarf candidates from microlensing with OGLE-III. *Monthly Notices of the Royal Astronomical Society* 458: 3012–3026. doi:10.1093/mnras/stw426.

Who's Moon

Elizabeth R. Stanway

The Moon has been an object of fascination since human culture first began. The dominant presence of Earth's satellite in the night sky, its regular phases and relatively frequent eclipses make it a potent symbol of the unknown. The earliest science fiction involved tales of travel to the Moon, from the political satire of Lucian of Samosata in the second century CE, through to scientific romances at the end of the nineteenth (Verne 1865, 1869; Wells 1901). Long before the Space Race offered the prospect of turning science fiction to science fact, the Moon was inextricably linked in the popular mind with visions of space travel.

Doctor Who arrived on our screens in 1963, set against a cultural milieu defined in large part by the Space Race between the USSR and the United States of America. The speeches of American president John F. Kennedy, in particular his famous address at Rice University (Kennedy 1962), have been likened to those of a suitor offering the Moon as a love token to their beloved (Launius 2012). It was a gift that Western culture, including the television viewers of the United Kingdom, needed to restore their confidence after the perceived humiliation of early Soviet success with spaceflight (Lule 1991; Barnett 2013). Kennedy's assassination, just one day before *Doctor Who*'s first broadcast, raised his stated objective for the Space Race to a national obsession in the United States and gave its achievement the status of a task undertaken *in memoriam*.

Unsurprisingly then, the Moon is a recurrent symbol in *Doctor Who* history, appearing as a setting or focus in stories from the mid-1960s through to the 2010s. This essay explores the conceptual representations and role of Earth's Moon in *Doctor Who*[1] and discusses how they reflect the changing contemporary understanding of the satellite and its origin.

As we stand on the brink of a new Space Age, I evaluate how the prominence of the Moon in the *Doctor Who* universe reflects its wider cultural and political significance, and how its representation is informed by popular attitudes to space. In this context, I discuss issues of ownership and sovereignty, lunar exploitation, the dark side of the Moon, the Moon as a scientific space, and the waning of the Moon as Mars waxes instead at the new frontier.

Ownership and Sovereignty

The question of who owns the Moon has challenged philosophers and politicians alike. Is the Moon merely another territory to be conquered by the most powerful nation? Unlike territory on Earth, the challenges of reaching the lunar surface requires technical rather than military superiority. The 1960s Space Race involved the exercise of both—advanced technology driven by political ideology and under tight military control. As early as 1962, President Kennedy had warned of seeing the Moon "under a hostile flag of conquest" without ruling out the possibility of planting his own flag (Kennedy 1962). When the Earth's satellite was eventually reached in 1969, the claims made in this respect were carefully ambiguous: while an American flag was planted, the Apollo 11 astronauts, all military officers, made the claim "We come in peace for all mankind" rather than simply on behalf of the USA.

Representations of the Moon's sovereignty in *Doctor Who* stories from the 1960s are actually less mixed than the message sent by the Apollo 11 astronauts. The first three stories in which the Moon featured prominently all treated the Moon as a symbol of international cooperation. In "The Tenth Planet" (1966) missions to the Moon feature a West Indian astronaut, under the command of an American general, answerable to International Space Command. Similarly in "The Moonbase" (1966), which appeared soon afterwards, both the purpose (world weather control) and the personnel of the station are deliberately global. Strong accents and actors with a range of ethnicities make a clear statement that no one nation has sovereignty on the Moon. While the actors in "The Seeds of Death" (1969) are less diverse in terms of skin color and background, the use of the Moon is still transnational: now in the role of a transportation hub transferring people and material between cities. By "Frontier in Space" (1973) the Moon's usage has changed, becoming a penal colony, but still under the auspices of a world government.

The principle of restricting space exploitation to peaceful purposes was articulated as early as 1958 (NASA 1958; United Nations 1962), nonetheless the Space Race had placed it clearly in the realm of disputed territory to be secured. The panic that greeted the USSR's launch of the Sputnik orbiter in 1957 reflected a very real fear that by reaching space, the Soviet Union was gaining a tactical advantage (Lule 1991; Barnett 2013). It is perhaps remarkable that through the 1960s, while the rhetoric of the Space Race focused on ideology and supremacy, *Doctor Who* was anticipating—and implicitly advocating—a principle that only become enshrined in international law with the Outer Space Treaty (United Nations 1967). This agreement, to which the United Kingdom, United States, China and the USSR (succeeded by Russia) are all signatories, disallows territorial claims beyond Earth's atmosphere, exploitation of extra-terrestrial resources by any one nation, or the military use of space. The subsequent Moon Treaty (United Nations 1979) was more explicit still: lunar resources are the "common heritage of mankind" (article 11) and "exploration and use of the moon shall be the province of all mankind and shall be carried out for the benefit and in the interests of all countries, irrespective of their degree of economic or scientific development" (article 4). It is noteworthy that this treaty was not ratified by the United States, United Kingdom or any other major space power.

If, then, the Moon does not belong simply to one nation can we categorically state an alternative? Does *Doctor Who*'s Moon belong equally to humanity of all nations? Certainly, this is the case, as seen on screen, in the classic series. However, the picture elsewhere in the *Doctor Who* Universe, in particular that emerging since the turn of the twenty-first century, is somewhat more mixed.

The Fifth Doctor novel *Imperial Moon* (Bulis, 2000) presents a rare example of *Doctor Who* confronting the issue of lunar sovereignty directly. The novel, in which a military officer leads a secret nineteenth-century British Royal Navy moon landing, employs the steampunk aesthetic and Victorian sensibilities to add color to what might otherwise be a straightforward (albeit domed) jungle-planet story. Nonetheless, one of the first actions of the mission is planting the Union flag and claiming the Moon for Queen and Country. This provides a context in which the officer in command can question his role as an Imperial trailblazer—although even then the question is not whether Britain above all other humanity can claim the Moon, but rather whether hypothesized native life might challenge the claim. The narrative here is primarily one of race, gender roles and imperialism, but both the rectitude and the difficulty of establishing territorial claims in the Moon's hostile environment are explored through the ultimate failure of the mission.

An unusual use of the Moon to explore issues of sovereignty occurs in the two-part Eleventh Doctor story, "The Impossible Astronaut/Day of the Moon" (2011). This narrative initially presents the Apollo program in a manner that subtly perverts its established symbolism: rather than standing as an exemplar of human courage, ingenuity and determination, it is shown to result merely from an alien requirement for invention of a space suit. Common human ownership not just of the Moon but of their own planet, Earth, is challenged, and shown to be illusory. It is in large part due to the effective symbolism of the Apollo program that the resolution to this story—in which the Doctor makes use of the "One Small Step" moonwalk imagery to return humanity to independent control of their planet—is both emotionally and intellectually satisfying. As Neil Armstrong asserts humanity's ascendency over the Moon "in peace for all mankind," every other human grasps the understanding—and weapons—required to achieve the same on Earth. An unspoken subtext is that the Doctor has not only reclaimed the Earth for humans, but also reclaimed the Moon and the Apollo landings as symbols of human achievement and sovereignty.

With military spy satellites and programs such as the "Star Wars" Strategic Defense Initiative undermining one of the prime tenets of the Outer Space Treaty through the 1980s and 1990s, it is perhaps unsurprising that representations of the Moon's sovereign status have also become less clear-cut over time. The children's novel *Dust of Ages* (Richards 2009) and television story "Kill the Moon" (2014) both suggest the Moon is being opened up for commercial exploitation, while the Eleventh Doctor novel *Apollo 23* (Richards 2010) and the audio adventure "1963: The Space Race" (Big Finish Productions 2013) both posit an American military base concealed on the lunar surface. While lunar mining may plausibly be permitted by the Outer Space Treaty, a permanent military presence outside Earth's atmosphere certainly is not.

Perhaps the most surprising conception of lunar sovereignty appearing in the series is in "Smith and Jones" (2007). In removing a London hospital and its occupants to the Moon, the alien Judoon are taking it to *neutral* territory. Far from recognizing the rights to lunar sovereignty of any nation, or even of humanity itself, the Shadow Proclamation (the strongest judicial authority in *Doctor Who* at this epoch) considers it outside of Earth jurisdiction entirely. Surprisingly, neither the Doctor nor any human challenges this interpretation. It is perhaps indicative of the popular indifference to the Moon in the early twenty-first century that this usurpation of "the common heritage of mankind" passed without comment.

The Moon as a Resource

While the question of lunar sovereignty has taxed philosophers, a more practical issue since the dawn of the Space Age has been the question of the Moon's resources, and how to use them. The Moon Treaty is clear that any profit derived from lunar resources must be shared with other nation states, regardless of the funding source for exploitation—however this document was never ratified by any space power, leaving the current position unclear (Hoffstadt 1994).

Commercial mining of the lunar regolith (the fine powdered rock coating the Moon's surface) could plausibly harvest materials including Helium 3, an important potential power source, as shown in the blockbuster film *Moon* (2009). However uncertainties regarding lunar ownership rights act as a discouragement for both government and private investment, since a return on investment is difficult to guarantee (Bilder 2010; Hertzfeld and von der Dunk 2005). Nonetheless, *Doctor Who*'s Moon is not immune to commercial pressures. A privately financed Mexican mineral survey is referenced in "Kill the Moon" and mineral prospecting on behalf of "the big corporations" in *Dust of Ages*. However, despite these examples, lunar mineral and resource exploitation has not been a major feature of *Doctor Who*, perhaps due to the surfeit of atmosphere-rich worlds, which might be more readily exploited.

Instead, the key practical utility of the Moon in the *Doctor Who* universe has been its strategic location. This is demonstrated, as already mentioned, by its use as a transportation hub (in "The Seeds of Death") and as a weather control station (in "The Moonbase"). In both cases, the application relies on a crucial feature of the Moon-Earth system: tidal locking. Gravitational forces act to transfer angular momentum from spin to orbit over time, until first the less massive object in an orbiting binary and then the more massive object come to synchronize, with one face constantly turned towards its companion. The Moon is tidally locked to the Earth. The Earth is yet to lock with the Moon, but is gradually slowing in its rotation as energy is dissipated, raising tides in both ocean and rock (Halley 1695; Stephenson and Morrison 1995). This property means that a site on the near side of the Moon will at all times have an entire Earth hemisphere in view. Thus, the Moon forms a natural communications satellite, and the writers of the 1960s serials used it as such. The rise of commercially viable artificial satellites, many in more useful geosynchronous orbits, rendered the Moon's use as an Earth-observation and communications platform redundant, and this is reflected in the absence of such stories from the *Doctor Who* canon since the 1960s.

The Dark Side of the Moon

The tidal locking of the Moon also results in a lunar hemisphere, which is never seen from the Earth's surface. This unobservable region has received the popular label "the dark side," where the misleading term refers to mystery rather than physical illumination (both sides of the Moon partake equally in its 709-hour day/night cycle). A similar dark side exists in the cultural symbolism of the Moon, with its influence blamed for afflictions ranging from lunacy to lycanthropy. Inevitably then, there is also a dark side to *Doctor Who*'s Moon.

The Moon itself has presented a threat to life on Earth in the *Doctor Who* universe. In "The Silurians" (1970), the eponymous reptiles retreated into underground hibernation through fear that an approaching minor planet would strip the atmosphere from Earth, rendering it uninhabitable. Instead, the planetoid was captured in a terrestrial orbit, becoming the Moon. At the time the "capture hypothesis" for the Moon's origin was first discussed on screen, the notion was widely accepted in scientific circles (Urey 1966). It was not until chemical analyses of the Apollo and Russian Luna mission moon rock samples were published that the modern "giant impact hypothesis" (Hartmann and Davis 1975, Brush 1988) replaced it. The common chemical composition of the Earth's crust and the lunar regolith requires a shared origin for the two bodies that the capture hypothesis cannot explain. It is noteworthy that, despite these scientific advances, capture has been retained as the canonical origin for the Moon in the *Doctor Who* universe, referenced in "The Hungry Earth/Cold Blood" (2009), "Dinosaurs on a Spaceship" (2012) and "Kill the Moon."

This last episode also features a second direct danger from the Moon: gravitational disturbances and consequent natural disasters that result from an embryo attempting to hatch from the lunar interior. Given that an egg is a closed system unable to gain mass, it is difficult to find a plausible scientific interpretation of this story, as discussed later. A more folkloric but similarly implausible threat from the Moon is seen in the ability of moonlight to activate lycanthropic alien cells in "Tooth and Claw" (2006) and, in concentrated form, to destroy them. The light of the full Moon is weakened, reflected sunlight from which the blue—highest energy, and most biologically active—wavelengths have been scattered (Jones et al. 2013). Hence, there is no known mechanism by which lunar illumination can excite intense biological activity left unaffected by sunlight.

A less direct threat, the use of the Moon as a beachhead or shield for alien invasion, is a frequent occurrence in the *Doctor Who* universe. The lunar surface provides a convenient location both for Earth observation and for influencing its atmosphere (as was seen in a more positive context above). The tidal locking of the Moon also leads to almost an entire hemisphere that is unobservable from Earth, but sufficiently close that any base or spacecraft sited there could receive signals from the Earth's surface. These would have to be bounced off small, easily overlooked communication satellites in lunar orbit, but allow a full-sized spacecraft to remain hidden. Instances of alien invasions launched from on or behind the Moon appear in stories including "The Moonbase," "The Invasion" (1968), "The Seeds of Death," "Attack of the Cybermen" (1985), "Silver Nemesis" (1988) and "Victory of the Daleks" (2010).[2]

The use of Earth's Moon as a site for disposal of waste products is another familiar trope from science fiction explored in the *Doctor Who* universe. Such a usage is certainly against the spirit, if not the language, of the Outer Space and Moon Treaties. Nonetheless, both the technical aspects (Burns et al. 1978) and the legal ones (Dusek 1997) have been extensively discussed. If the delivery mechanism could be rendered safe then the remoteness of the lunar surface and its absence of life offer particular advantages for nuclear waste disposal.[3] While this particular usage does not feature explicitly in *Doctor Who*, a parallel disposal problem—that of human "waste" in the form of convicted criminals or political prisoners—is a prominent feature of the serial "Frontier in Space." Here the Moon appears as a penal colony (as it also does in "Bad Wolf" [2005] and the novel *Apollo 23*), fortified by the poor survival chances of escapees leaving the prison's controlled environment. While this concept appears elsewhere in science fiction (Heinlein 1968, Baxter 2015), it attracts little serious discussion outside of the fictional realm. The prohibitive cost of such a facility and of prisoner transport cannot be justified in any realistic scenario (Jones 1992).

The Moon as a Scientific Space

Unlike in other, contemporary science fiction (e.g., *Captain Scarlet*, 1969) civilian lunar colonization has never been a major feature of the *Doctor Who* universe. A rare example is perhaps implicit in "Let's Kill Hitler" (2011), where a character refers to having studied at Luna University. A university on the lunar surface continues a narrative that positions the Moon as a space for science, rather than for military, commercial and utilitarian purposes. The Apollo missions allocated time and weight to a number of experiments carried out *in situ* on the lunar surface and placed a lunar-ranging reflector, which is still in use. The Moon also offers a powerful site for astronomical telescopes. The 15cm diameter ultraviolet telescope carried by the Chinese Chang'e 3 moon lander produced cutting-edge scientific observations (Zhu et al. 2016; Wang et al. 2015). Its small size is offset by the stable platform outside of Earth's ultraviolet-opaque atmosphere, which enabled investigations simply not possible from the Earth's surface.

While not explicitly used as an astronomical observatory, the Moon in *Doctor Who* is very much a scientific space. Both the "T-Mat" transmat hub in "The Seeds of Death" and the Gravitron weather-control device in "The Moonbase" are complex devices, with support crews of both technical staff and pure scientists. Indeed, the Second Doctor and his companions are explicitly introduced to physicists, astronomers, mathematicians and geologists (although curiously no meteorologists) in "The Moonbase," and the premise on which the Gravitron itself operates is clearly stated: "The Gravitron controls the tides. The tides control the weather. Simple, eh?" The same story includes both visual and narrative discussion of conditions on the lunar surface, including its surface gravity, the need for breathing equipment and eye protection, atmosphere-controlled and domed human environments, the effects of the long lunar day on the human biological clock, the Doppler effect and the chemistry and effect of solvents such as acetone. It is surely no coincidence that the story's writer was a professional scientist: ophthalmologist Christopher "Kit" Pedler. However, the willingness of the production team to commission and produce a script heavy in technical detail also reflects the often-discussed educational remit of *Doctor Who*. Nor was this an isolated example. The underlying physics behind the T-Mat in "The Seeds of Death" is less apparent than that of the Gravitron, but the story is still rich in technical terminology and prominent technician characters. The Moon in this serial provides a relatively rare setting in which companion Zoe can demonstrate her own credentials as a scientist and—a still rarer occurrence—interact with another female scientist on a technical level without intervention from male colleagues. While such conversations might plausibly have occurred in any Earth-based or human-settlement context, the isolated and technology-dependent artificial environment of the Moon provides a natural space for exploring these scientific interactions.

Not only is the Moon one of the most striking astronomical phenomena in Earth's sky, its tides have a pronounced effect on Earth itself. As a result, the process of learning about the Moon is integral to learning about Earth, and forms part of the national curriculum for primary school children in the UK (Department for Education 2015). In this context, and with imagery of the US and Russian space programs filling television screens, it is perhaps unsurprising that the early *Doctor Who* stories took pains to make their representation of the Moon plausible and factually accurate (insofar as possible within the science fiction context). This approach was particularly explicit in the show's early years, but never entirely abandoned. Both factual and speculative articles on real-world spaceflight were regular features in the *Doctor Who* annuals published through the 1960s, 70s and 80s (e.g., BBC 1967, 1976, 1978, 1982) and have continued since the series re-launch in 2005. From the inclusion of factual "Tardis Data Bank" entries in the children's book *Dust of Ages,* to an account of "The Real Apollo Missions" in the *Doctor Who Official Annual 2012* (BBC 2011), scientifically accurate information is important when the Moon is discussed. Overtly incorrect representation of the Moon would be apparent even to young viewers and would jar the requisite suspension of disbelief.

In this context, it is impossible to overlook the one striking counter-scientific example in recent years. "Kill the Moon" sets up an initially plausible scenario, invoking both commercial ore prospecting on the Moon and the concept that changes in the Moon's gravity are causing natural disasters on Earth. However, from that point onwards, it is recklessly indifferent to scientific fact, demonstrating serious misconceptions in areas ranging from microbiology to gravitational astrophysics (see Plait 2014). In part, this may reflect the writer's focus on emotional aspects of the script and its spotlight on developing the Doctor-Clara relationship, but it is still indicative of a decline in the quality of scientific and educational representation of the Moon.

Research shows *Doctor Who* has influenced the decision-making and attitudes of its fans on scientific issues (Orthia 2019). Perhaps surprisingly, there is little evidence that *Doctor Who* episodes set on the Moon, or indeed Mars, inspired a wider interest amongst the public in the United Kingdom. The figure on the following page makes use of the Google Trends analytic tool to evaluate whether the first broadcast of episodes that feature either world has led to any measurable increases in web browser searches for the terms "Moon" or "Mars," using the ubiquitous Google search engine.[4] The analysis is limited to searches from the UK, in order to avoid confusion due to delayed broadcasts in other regions, and to first-time broadcasts since the re-launch of *Doctor Who* in 2005. The date of key events and press releases concerning these worlds are also indicated. As the figure demonstrates, there is no correlation between *Doctor Who* broadcasts and web searches. Indeed the only clear relation between events and searches is seen in two merged peaks dating from 2009, which mark the release of the Hollywood blockbuster *Moon* (2009), and the announcement that water ice had been detected on the lunar surface.

The Waning Moon?

If, then, the Moon is such a prominent potential location, and such an important symbol in many different domains, why is it not used more in the *Doctor Who* Universe? The Second Doctor era of 1966–1969 featured three major stories in as many years in which the plot revolved around either a moon base or a mission to the Moon, and another in which the

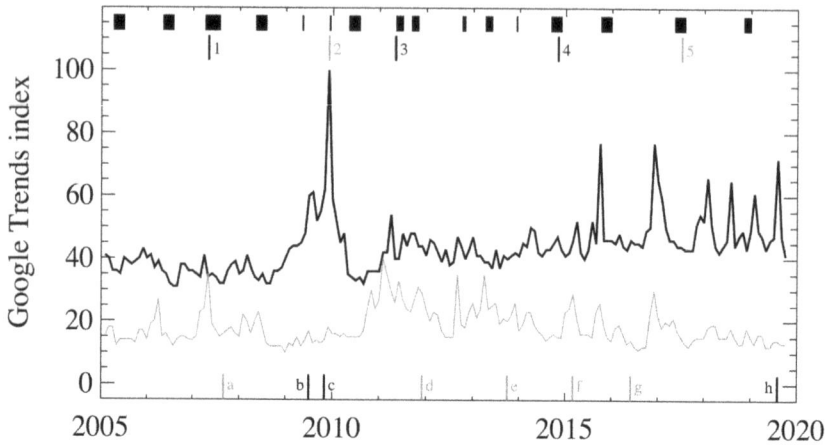

Google Trends analytics for UK searches for the terms "Moon" (black) and "Mars" (gray) over the period Jan. 2004–Sep. 2019. Data indicate the number of Google web searches per month, scaled such to a peak of 100. UK first transmission dates for *Doctor Who* are indicated by black bars, together with dates for five specific stories: (1) "Smith and Jones," (2) "The Waters of Mars," (3) "The Impossible Astronaut/Day of the Moon," (4) "Kill the Moon" and (5) "Empress of Mars" (2017). Selected real-world events are indicated above the date axis: (a) Landing of the Phoenix Rover on Mars, (b) opening of 2009 movie *Moon*, (c) announcement of water on Moon detected by LCROSS, (d) landing of Curiosity rover on Mars, (e+f) recruitment announcements by Mars One crowdfunded program, (g) SpaceX announcement of Mars plans, (h) Apollo 11 50th anniversary.

Moon shielded invasion. By contrast, 1970s *Doctor Who* included only one story in which the Moon appeared—and then in the context of a prison planet, which might plausibly have been relocated elsewhere in the Solar System. Despite the overwhelmingly Earth-based stories of the Third Doctor era and into the Fourth, there is an increasing prominence given to stories featuring Mars, rather than the Moon, with entries including "The Ambassadors of Death" (1970) and "Pyramids of Mars" (1975). This trend continued through the prose and audio fiction of the 1990s and 2000s, with stories involving Mars far outnumbering those involving lunar settlement or exploration.

This was likely a practical and story-driven decision. The Apollo missions had drawn a firm line under the already long-discredited theory that there might be life on the lunar surface. As such, any adventure set there must necessarily involve interactions between human beings rather than between humans and aliens, unless the aliens themselves were already displaced. At the same time, the grey landscape of the Moon offered little visual excitement on televisions newly converted to color. Instead, alternate locations on Earth—or hypothesized Earth-like worlds (see Eldridge, this volume)—provided scope for more interesting settings, and removed the necessity for expensive set designs and contrived explanations for Earth-like gravity.

However, this shift also reflects a similar change in emphasis in military and geopolitics, as well as in the popular support—or lack thereof—for human-crewed space exploration. In both the United Kingdom and the United States, "after the first Moon landing, the public rapidly became bored with the technology of space travel, and it needed a human drama such as the problems of Apollo 13 to engage their interest" (Jones 2004: 47). Support for NASA and its (very expensive) work plummeted against a backdrop of financial hardship and an easing of the Cold War in the 1970s (Roy et al. 2000; Launius 2003). It rapidly became clear that neither military nor commercial infrastructure on the Moon was financially viable at the then-current technology level. Despite the substantial scientific and financial benefits accrued from Apollo program research, there was a perception that the Moon landings had done little of value beyond their initial publicity. While a majority of Americans supported the concept of further crewed space flight, a similar majority believed that the requisite money could be better used (Roy et al. 2000). At the same time, probes visiting Venus and Mars raised their public profile, with the latter at least considered a promising location for both alien life and human settlement. Given the impracticality of carrying out interplanetary missions on the post–Apollo NASA (or Russian space effort) budget, crewed spaceflight refocused on low Earth orbit, with commercial exploitation reaching out further to geosynchronous orbit.

Largely, this paradigm has continued to the current time. As recently as 2008, only about 55 percent of respondents to a survey undertaken at a public scientific exhibition in the UK supported space exploration, with a fifth considering Mars a viable target, and no more than 2 percent supporting a return to the Moon (Entradas and Miller 2010). In an audience without an extant interest in science, the numbers would likely have been lower. As an astronaut, Lundvik, told the Doctor in "Kill the Moon": "We stopped going into space. Nobody cared."

Paradoxically, that very statement may itself be a consequence of a new and favorable shift in attitudes toward crewed space exploration. Successive American presidents since George Bush senior in 1989 have announced initiatives involving a return to the Moon and crewed missions to Mars (although without commensurate increases in NASA funding, and all but the most recent subsequently cancelled, Launius 2012). At the same time, the development of communications and information technology has improved the potential viability of commercial exploitation of space, and empowered a generation of technically-minded enthusiasts, entrepreneurs and philanthropists—many of whom were themselves raised on tales of the Moon landings. NASA, once the bastion of technology in the service of national interests, has engaged in a series of initiatives to recruit and encourage commercial alternatives (e.g., NASA 2019a, 2019b). Financial inducements have also been offered to encourage commercial lunar missions, primarily the Google Lunar X-Prize (XPrize Foundation 2019) which ran from 2007 to 2018 and generated programs that are still ongoing (Matthew et al. 2019). The drive towards commercial or crowd-funded space exploitation has been termed the New Space Age—or simply "New Space" (see Pomeroy et al. 2019 and references therein). This landscape of private investment and minimal government regulation plays an important role in twenty-first century discourses on space exploration and has the potential to open space to countries without substantial government funding.

Legislators and legal theorists are still racing to catch up with this commercial momentum and the complex legal issues it raises (Bilder 2010; Dempsey 2014; Bruhns and Haqq-Misra 2015). Nonetheless, this image of a commercialized, highly technological future for space has become prominent in science fiction. It characterizes the lunar mining operation in *Moon* (2009), and is at the root of the space debris removal attempt in *Gravity* (2013). Another example, closer to home for *Doctor Who*, can be found in the BBC's 2005 adaptation of *The Quatermass Experiment* (originally broadcast in 1953) in which the British Experimental Rocket Group is transformed from its origins as a government organization to a commercial entity for the twenty-first century.

While the enthusiasm over unrealistic early crowd-funding efforts (notably the Mars-One program) has ebbed (Slobadian 2015, O'Callaghan 2019), the New Space Age is far from moribund. NASA's efforts include design studies for the Project Gateway lunar-orbital space station, and NASA and ESA have recently exhibited their completed crew capsule for the joint, moon-targeted Artemis program. At the same time, China has landed successful rovers and probes on the lunar surface, and India came very close to achieving a soft lunar landing in the summer of 2019.

The New Space vision has been reflected in the renewed significance of both the Moon and Mars to the *Doctor Who* universe since its return in 2005. Examples of commercial exploitation of the Moon have already been discussed, while "The Waters of Mars" (2009) references both national and commercial efforts to reach the red planet.[5] In this new paradigm, the Moon is seen as a test-bed and stepping-stone for future missions to Mars, and a vision expressly articulated by *Doctor Who* in the biography of base commander Adelaide Brooke ("The Waters of Mars"). The Moon itself is no longer the primary objective. Short of a "Kill the Moon"–scale spectacle, the power of our Moon to capture imaginations and attract resources is, it seems, on the wane.

Nonetheless, in the epoch of the New Space Age, *Doctor Who* is there to inform, educate and entertain, just as it was in the first Space Age. As such, a return, both to the Moon and Mars, is probable in the near future of the Doctor, and appears increasingly possible even in the world that we, as viewers, inhabit.

Notes

1. An exhaustive list of references to the Moon in *Doctor Who* on screen, in literature, audio, comics, annuals and other media is likely impossible to construct. Instead, illustrative examples will be discussed where appropriate.

2. Four of these examples involve attacks by the Cybermen who seem irresistibly attracted to Earth's Moon—perhaps because its use as a barrier to observation mirrors the Sun's role in hiding Mondas from Earth, or perhaps because the otherwise-twinned world of Mondas lacks a large natural satellite comparable to Earth's.

3. A balance-of-bad-alternatives judgment that provides the premise for the television series *Space:1999* (1975–77).

4. Google Trends assigns a score to the search frequency in each calendar month since 2004, scaled such that the maximum monthly rate in any given search (or comparison between searches) scores 100. See https://trends.google.com and discussion in Whitman Cobb (2015).

5. The audio drama "Red Dawn" (Big Finish Productions 2000) represents a relatively early example of this trope.

References

Baxter, S. (2015) Governance of a Free Moon in Science Fiction. In: Cockrell, C.S. (ed.) *Human Governance Beyond Earth*. Springer, pp. 63–80.
BBC (1967) *The Dr Who Annual 1968*. World Distributors.
BBC (1976) *The Dr Who Annual 1977*. World Distributors (Manchester) Ltd.
BBC (1978) *The Dr Who Annual 1979*. World Distributors (Manchester) Ltd.
BBC (1982) *BBC TV Doctor Who Annual*. World International Publishing Ltd.
BBC (2011) *Doctor Who: The Official Annual 2012*. Penguin Books.
Bilder, R.B. (2010) A Legal Regime for the Mining of Helium-3 on the Moon: U.S. Policy Options. *Fordham International Law Journal* 33: 243–299.
Bruhns, S., and Haqq-Misra, J. (2016). A Pragmatic Approach to Sovereignty on Mars. *Space Policy* 38: 57–63. doi:10.1016/j.spacepol.2016.05.008.
Brush, S.G. (1988) A history of modern selenogony: Theoretical origins of the Moon, from capture to crash 1955–1984. *Space Science Reviews* 47: 211–273. doi:10.1007/BF00243556.
Bulis, C. (2000) *Doctor Who: Imperial Moon*. BBC Worldwide Ltd.
Burns, R.E., Causey, W.E., Galloway, W.E., and Nelson, R.W. (1978) Nuclear Waste Disposal in Space. *NASA Technical Paper 1225*.
Dempsey, P.S. (2014) National Legislation Governing Commercial Space Activities. *Journal of Space Safety Engineering* 1: 44–60.
Department for Education (2015) National curriculum in England: Science programmes of study. www.gov.uk/government/publications/national-curriculum-in-england-science-programmes-of-study (accessed 6 Oct 2019).
Dusek, R. (1997) Lost in space: The legal feasibility of nuclear waste disposal in outer space. *William and Mary Environmental Law and Policy Review* 22: 181–218.

Entradas, M., and Miller, S. (2010) Investigating Public Space Exploration Support in the UK. *Acta Astronautica* 67: 957–953. doi:10.1016/j.actaastro.2010.06.015.

Halley, E. (1695) Some Account of the Ancient State of the City of Palmyra, with Short Remarks upon the Inscriptions Found there. *Philosophical Transactions of the Royal Society* 19: 160–175. doi:rstl.1695.0023.

Hartmann, W.K., and Davis, D.R. (1975) Satellite-Sized Planetesimals and Lunar Origin. *Icarus* 24: 504–515. doi:10.1016/0019-1035(75)90070-6.

Heinlein, R.A. (1968). The moon is a harsh mistress. New York: G.P Putnam's Sons.

Hertzfeld, H.R., and von der Dunk, F. (2005) Bringing Space Law into the Commercial World. *Chicago Journal of International Law* 6: 81–99.

Hoffstadt, B.M. (1994) Moving the Heavens: Lunar Mining and the Common Heritage of Mankind in the Moon Treaty. *UCLA Law Review* 42: 575–622.

Jones, A., Noll, S., Kausch, W., Szyszka, C., and Kimeswenger, S. (2013) An advanced scattered moonlight model for Cerro Paranal. *Astronomy & Astrophysics* 560: A91 (11pp). doi: 10.1051/0004-6361/201322433.

Jones, E.M. (1992) A basis of settlement: Economic foundations of permanent pioneer communities, *The Second Conference on Lunar Bases and Space Activities of the 21st Century* 2: 697–701.

Jones, R.A. (2004) They came in peace for all mankind: Popular culture as a reflection of public attitudes to space. *Space Policy* 20: 45–48. doi:10.1016/j.spacepol.2003.12.002.

Kennedy, J.F. (1962) Address at Rice University, 12 September 1962. www.jfklibrary.org/asset-viewer/archives/USG/USG-15-29-2/USG-15-29-2.

Launius, R.D. (2003) Public Opinion Polls and Perceptions of, U.S. Human Spaceflight. *Space Policy* 19: 163–175. doi:10.1016/S0265-9646(03)00039-0.

Launius, R.D. (2012) Why go to the Moon? The many faces of lunar policy. *Acta Astronautica* 70: 165–175. doi:10.1016/j.actaastro.2011.07.013.

Lule, J. (1991) Roots of the Space Race: Sputnik and the Language of U.S. News in 1957. *Journalism Quarterly* 68: 76–86. doi:10.1177/107769909106800109.

Matthew, J., Nair, B.G., Safonova, M., Sriram, S., Prakash, A., Sarpotdar, M., Ambily, S., Nirmal, K., Sreejith, A.G., Murthy, J., Kamath, P.U., Kathiravan, S., Prasad, B.R., Brosch, N., Kappelmann, N., Gadde, N.S., and Narayan, R. (2019) Prospect for UV observations from the Moon. III. Assembly and ground calibration of Lunar Ultraviolet Cosmic Imager (LUCI). *Astrophysics and Space Science* 364: A53. doi:10.1007/s10509-019-3538-8.

National Aeronautics and Space Administration (NASA) (1958) National Aeronautics and Space Act of 1958, *Public Law #85–568, 72 Stat., 426.* history.nasa.gov/spaceact.html

National Aeronautics and Space Administration (NASA) (2019a, June 20). Commercial Lunar Payload Services (CLPS) On-Ramping. *Solicitation Number: 80JSC019R0013.*

National Aeronautics and Space Administration (NASA) (2019b, July 30). NASA Announces Call for Next Phase of Commercial Lunar Payload Services. Press Release ID 19–064. www.nasa.gov/press-release/nasa-announces-call-for-next-phase-of-commercial-lunar-payload-services.

O'Callaghan, J. (2019, February 11). Goodbye Mars One, The Fake Mission to Mars That Fooled the World. *Forbes.* www.forbes.com/sites/jonathanocallaghan/2019/02/11/goodbye-mars-one-the-fake-mission-to-mars-that-fooled-the-world/.

Orthia, L.A. (2019) How does science fiction television shape fans' relationships to science? Results from a survey of 575 *Doctor Who* viewers. *JCOM* 18: A08. doi:10.22323/2.18040208.

Plait, P. (2014, 6 October) Slate Plus *Doctor Who* Podcast Episode 7: "Kill the Moon." *Bad Astronomy.* slate.com/technology/2014/10/slate-plus-doctor-who-podcast-episode-7-kill-the-moon.html (accessed 6 Oct 2019)

Pomeroy, C., Calzada-Diaz, A., and Bielicki, D. (2019) Fund Me to the Moon: Crowdfunding and the New Space Economy. *Space Policy* 47: 44–50. doi:10.1016/j.spacepol.2018.05.005.

Richards, J. (2009) *Doctor Who: Dust of Ages.* BBC Children's Books: Penguin.

Richards, J. (2010) *Doctor Who: Apollo 23.* BBC Books: Ebury Publishing.

Roy, S.A., Gresham, E.C., and Christensen, C.B.(2000). The Complex Fabric of Public Opinion on Space. *Acta Astronautica* 47: 665–675. doi:10.1016/S0094-5765(00)00104-1.

Slobodian, R.E. (2015) Selling space colonization and immortality: A psychosocial,

anthropological critique of the rush to colonize Mars. *Acta Astronautica* 113: 89–104. doi:10.1016/j.actaastro.2015.03.027.
Stephenson, F., and Morrison, L. (1995) Long-Term Fluctuations in the Earth's Rotation: 700 BC to AD 1990. *Proceedings of the Royal Society A* 351: 165–202. doi:10.1098/rspa.2016.0404.
United Nations (1962) Declaration of Legal Principles Governing the Activities of States in the Exploration and Use of Outer Space. *General Assembly resolution 1962 (XVIII)*. www.unoosa.org/oosa/en/ourwork/spacelaw/principles/legal-principles.html.
United Nations (1967) Treaty on Principles Governing the Activities of States in the Exploration and Use of Outer Space, including the Moon and Other Celestial Bodies. *General Assembly resolution 2222 (XXI)*. www.unoosa.org/oosa/en/ourwork/spacelaw/treaties/outerspacetreaty.html.
United Nations (1979) Agreement Governing the Activities of States on the Moon and Other Celestial Bodies. *General Assembly resolution 34/68*. www.unoosa.org/oosa/en/ourwork/spacelaw/treaties/moon-agreement.html.
Urey, H.C. (1966). Chemical evidence relative to the origin of the solar system. *Monthly Notices of the Royal Astronomical Society* 131: 199–223. doi:10.1093/mnras/131.2.199.
Verne, J. (1865). *De la terre à la lune*. Paris.
Verne, J. (1869). *Autour de la lune*. Paris.
Wang, J., Wu, C., Qiu, Y.L., Meng, X.M., Cai, H.B., Cao, L., Deng, J.S., Han, X.H., and Wei, J.Y. (2015) An unprecedented constraint on water content in the sunlit lunar exosphere seen by Lunar-based Ultraviolet Telescope of Chang☒e-3 mission. *Planetary and Space Science* 109–110: 123–128. doi:10.1016/j.pss.2015.02.006.
Wells, H.G. (1901) *The First Men in the Moon*. London: GW Newnes Ltd.
Whitman Cobb, W.N. (2015) Using big data to examine public opinion of space policy. *Space Policy* 32: 11–16. doi:10.1016/j.spacepol.2015.02.008.
XPrize Foundation (2019). Google Lunar XPrize: The New Space Race. lunar.xprize.org (accessed Oct 7 2019).
Zhu, L.Y., Zhou, X., Hu, J.Y., Quan, S.B., Li, L.J., Liao, W.P., Tian, X.M., and Wang, Z.H. (2016) LUT reveals an Algol-type eclipsing binary with three additional stellar companions in a multiple system. *Astronomical Journal* 151: 107 doi:10.3847/0004-6256/151/4/107.

$E=mc^3$

Doctor Who *and Energy*

Marcus K. Harmes

Introduction: All Energy Is a Source of Life

Many may know Albert Einstein's equation $E=mc^2$, a statement of the equivalence of mass and energy. However, the 1972 *Doctor Who* story "The Time Monster" turns the familiarity of this formula on its head, when the alien criminal the Master declares that energy equals mass times the speed of light *cubed*, at least in the field of "extra-temporal physics of the time vortex." The equation $E=mc^3$ and the Master's scientific explanation may be just so much gobbledegook, but it is part of the texture of a story preoccupied with energy and part of an era of the program which evidences the same preoccupation. The distortion of possibly the most famous equation in both scientific circles and among popular knowledge of science is also on trend with an era of *Doctor Who* which unsettled or perturbed understandings of energy and what it contributed to society. The actual equation $E=mc^2$ was cited in "The Dæmons" in 1971, to give a patina of scientific accuracy to the strange events happening on screen, including a heat wave and a sudden drop in temperatures. These the Doctor explains as the by-products of an alien spaceship shrinking and then expanding, for "when you lose mass the energy has to go somewhere." Even using the correct equation, the associations built around energy are uncanny and sinister.

Three years earlier, the color era of the show began with the serial "Spearhead from Space" (1970), in which a character, who is also an alien entity bent on the invasion of Earth and the destruction of humankind, declares, "All energy is a source of life." The statement is not neutral; in the context of the story, in which a hostile alien force exists purely as energy in a disembodied state, the energy has sinister overtones. The association of the sinister, the unsettling and the potentially threatening with energy in this dialogue is of a type across the program.

This essay examines the *Doctor Who* stories made up to circa 1980, stories (or serials) broadly contextualized by Britain's own crises with energy sources including coal and petrol. Concerns about energy resonated at a practical level: would the lights stay on? However, concerns reached existential levels about the end of the world. This essay examines serials that range in production timing from between the Cuban Missile Crisis (1962) to the so-called "Winter of Discontent" (1979), and therefore from fear that atomic energy would end the world, to concern at the failure of national energy and labor. Common to many serials in this period are scientific researches and researchers seeking new or alternative sources of energy, transferring energy from both "savage" and "advanced" people, or seeking putatively limitless sources of energy. Contextualized as it is by Britain's own struggles to stabilize national energy supplies, the essay examines how *Doctor Who* characterizes the sources of energy, the science creating them, and the scientists responsible. In a national context where energy was parlous and stable sources may well have been welcome and necessary, it is noteworthy therefore that the program often suggests the contrary, with sources of energy from the conventional to the alternative characterized as threats and their creators as irresponsible. The scholarship on modern science fiction has noted that science fiction will often prognosticate negatively. American science fiction took a notably negative turn during the 1950s, in which technological advance became something fearful rather than hopeful (Warrick 1980). This essay will build on from this point. It begins by considering the different ways *Doctor Who* characterizes and uses energy and how the contemporary scientific, industrial and political context gives meaning to the energy within the series. Then it will address the specific anxiety about energy that registers in *Doctor Who* that it would destroy the world.

Energy and the Origins of Doctor Who

Doctor Who is a fruitful source to understand how science filters into popular discourse, particularly energy. In the first story, "An Unearthly Child" (1963), the mystery of a strangely intelligent schoolchild and her reclusive grandfather turns on the subject of energy. The mysterious police box that is their home in a junkyard hums with power from an unseen source. Two teachers investigating their mysteriously intelligent pupil touch the police box and find "It's alive!" even though "It's not connected to anything." When the two teachers burst inside what they soon discover is not actually a police box at all but the TARDIS, and an argument with its pilot, the Doctor, then ensues. The Doctor refuses to explain how TARDIS works, insisting, it "doesn't roll along on wheels" but does possess the power of "free movement in time and space."

The show's first serial artfully contrasts the high technology with the most elemental. The gleaming electronic interior of the TARDIS, brought onto the screen as a pulsating and living technological environment via a mixture of set design, ambient sound, lighting and dialogue, contrasts with the filthy hovel of cave dwellers when the Doctor lands the TARDIS in the Stone Age. These cave dwellers seek the most basic form of energy, the ability to spark a flame and create fire, with the Doctor called on to play the promethean part of the fire maker.

The TARDIS's energy sources are described lyrically in the next story, "The Daleks" (1963–64), with the Doctor referring to its "philosophy of movement" when asked to account for its functioning. Other energy sources abound in the storytelling of early *Doctor Who*. Viewers and characters learn that the TARDIS relies on mercury in its fluid links, without which the ship will not function. The alien creatures the Daleks need static electricity, and characters evocatively describe their reliance on the contact between their casings and the metal floor and static electricity becomes essential to the plot. To evoke the idea of the static contact, a character comments, "Have you noticed, for example, that when they move about there's a sort of acrid smell," like the smell of dodgem cars at a fairground and the Doctor explains for both his companions and the watching audience, "the Daleks have discovered a way to exploit static electricity." To escape from captivity the Doctor and his companions use a nylon cloth to break the contact between the Dalek and the floor. The same use of science enabled the plot resolution in the 1966 serial "The Power of the Daleks" in which the Doctor defeated the titular enemies by switching off the electricity supply. Again the energy source is vividly characterized: "They're powered by static electricity. It's like blood to them, a constant life-stream." Back in the Daleks' first serial in 1963, there are other sources of energy in addition to the electric, that speak to the political environment surrounding the serial's creative context, as the previous year the Cuban Missile Crisis brought the United States and the USSR close to a tactical nuclear exchange (Norris and Kristensen 2012). The Daleks have reactors full of nuclear matter and they plan to bombard their world, the planet Skaro, with radioactive matter. Energy on a broader cosmic scale is in the next story, "The Edge of Destruction" (1964), when the TARDIS is traveling towards the birth of the solar system and a massive burst of energy. Again, the dialogue describes the energy evocatively and energy becomes intrinsic to the plot as the space-time craft suffers from mysterious power losses. The TARDIS has moved too near in time and space to the "force of a total solar system." The Doctor explains:

> We're at the very beginning, the new start of a solar system. Outside, the atoms are rushing towards each other. Fusing, coagulating, until minute little collections of matter are created. And so the process goes on, and on until dust is formed. Dust then becomes solid entity. A new birth, of a sun and its planets.

The intrusion of energy sources where they did not belong is a theme in "The Dalek Invasion of Earth" (1964), in which the Daleks have a gigantic mining operation underway. They intend to extract the Earth's core and replace it with a giant motor that would be "a power system that will enable us to pilot the planet anywhere in the universe." In the same year and season, "Planet of Giants" proposed that an energy source called "space pressure" had the capacity to shrink human beings. In these early serials therefore, sources of energy participated directly in the storytelling, being devices to drive plots and inspire action. The epic adventure "The Daleks' Master Plan" (1965–66) revolves in its entirety around the quest for a rare energy source, a "full emm of taranium." The energy source, as well as its unit of measurement, receives description in ways that endows it with uncanny properties and even a character of its own. The substance will power "the core of the time destructor," and has other sinister characteristics. The Doctor warns against even looking at it, for "It will burn your eyes. You'll go totally blind!" The emm of taranium in "The Daleks' Master Plan" is a rare instance of an energy source with a unit of measurement, whereas many other stories only vaguely describe and quantify the transfers of power between things or people.

These early stories introduced a focus on sources of energy that continued to register in the scripting and the plotting across the stories of the 1960s and into the 1970s. That is not to say that every serial evidences the same concerns or themes. *Doctor Who* is a patchwork of different creative impulses as production and writing teams changed every few years, and each serial had a different primary scriptwriter whose sources of inspiration varied widely (Harmes 2014). In the earlier seasons, some stories were also set in a historical period and devoid of science at all. However, as a program that identified itself primarily as science fiction, the serials returned regularly to energy. The types and sources varied, as did the intended functions. The next sections of the essay organize types of energy by associated traits, including their sources and implications.

The Conventional but Threatening

In serials when Patrick Troughton played the Doctor, energy continues to be a source of narrative urgency. "The Moonbase" (1967) is about a piece of weather-engineering equipment called the Gravitron, which "controls the tides, [because] the tides control the weather." It is mostly benign but is hijacked and misused. The notion that Earth scientists could control natural energy recurs in the serial "The Enemy of the World" (1967–68) when a would-be dictator has the energy to control the weather and uses it for corrupt purposes, possessing the technology to cause earthquakes and volcanoes at strategic times and places to destabilize governments. The antagonist, Salamander, has fabricated one disaster, a nuclear holocaust, to keep a team of technicians underground who operate the equipment creating actual disasters.

When *Doctor Who* became a color program in 1970, and with Jon Pertwee playing the Doctor, energy continued to be an animating creative force in stories. Energy that was recognizably plausible provided the basis of plotting. "The Silurians" (1970) is set in a giant underground energy plant where scientists claim they are "on the verge of discovering a way to provide cheap, safe, atomic energy for virtually every kind of use," one which "converts nuclear energy directly to electrical power." In common with other serials of Pertwee's first year, the writers ground the science fiction in a plausible reality of contemporary science around establishments and laboratories in southern England as well as workplace dynamics. In "The Silurians," abnormal and unsettling events center on an ancient reptile race and an out of control nuclear reactor as much as a female scientist's unplanned pregnancy and a spate of nervous breakdowns. Even though these quotidian workplace realities appear, they do not normalize the science but render it deeply sinister. While scientists experience their everyday personal dramas, underneath their base the prehistoric Silurians intend to hijack the nuclear reactor, convert its energy to microwaves, and use these to smash the Van Allen belt. Doing so will mean another source of energy, radiation, will destroy life on Earth. Meanwhile, the scientists in "The Silurians" were under pressure from their political masters in Whitehall to provide "Limitless supplies of cheap, safe energy."

The same pressure imposed by government underpins the narrative and science in "Inferno" (1970). A massive drilling operation and the scientist directing it are penetrating the Earth's crust to locate "a vast new storehouse of energy," believed to be a gas. As the narrative progresses, an energy source is indeed located and released, but far from being a beneficent and limitless energy supply, it is a source of destructive power that degrades humans into beasts and (in a parallel Earth the Doctor witnesses) bursts the planet asunder. In these stories, the energy carried certain characteristics. It was earthly not alien (even in "Inferno," the destructive forces unleashed by the drilling came from within the Earth with no suggestion of extraterrestrial origin) and the implication was simply wide scale destruction.

The pressure on science from the government to find cheap, domestic and limitless power drew off contemporary pressures. Two years after the broadcast of "The Silurians" and "Inferno," former National Coal Board chairman Lord Robens presented on long-standing schemes to make Britain "self sufficient for energy needs." His three-point plan to achieve that carried with it criticism of ministers and civil servants but also testified to concern at the heart of both government and industry that Britain was quite simply running out of power (*Birmingham Post* 20 November 1973, p.1).

The Alien and Uncanny

Doctor Who also imbued energy with sinister associations and characteristics. "The Evil of the Daleks" (1967) repeats the scientific associations in earlier serials of alien menace powered by static electricity. It extends that however to the human beings who have themselves been possessed by Dalek characteristics. The Doctor remarks of one character that, "Ever since I came to this house, I have never seen you eat or drink anything." Far from needing conventional human sustenance from food and drink, he is animated by a different power source, as he exerts a magnetic force on metal, and his mysterious powers and his uncanny freedom from normal human appetites are noted as being like something from an Edgar Allen Poe story. The association links the electric with the sinister. The energy generated by evil thought impulses in "The Dæmons" and the psychokinetic energy of the Mentiads in "The Pirate Planet" (1978) carry sinister associations, as does the energy gained from sucking on blood in "The Stones of Blood" (1978) and "State of Decay" (1980).

The Alien and Parasitic

The era of the program starring William Hartnell as the Doctor ended with serials about energy that characterized it as unnatural and parasitic, taking more than it gave. In "The Tenth Planet" (1966) a close twin, the planet Mondas, threatens the entire planet Earth. Not only does Mondas have its own propulsion system that enables it to travel through space, it also has the capacity to drain the Earth's energy, even the life force animating human beings. The serial's science fiction suggests the interrelationship between all types of energy. Mondas's inhabitants, the Cybermen, die when their own planet loses energy. On Earth, the energy drain affects humans as much as other sources of power.

In other serials, energy carried a different and more subtly dangerous character. In "The Claws of Axos" (1971) the energy was parasitic. What again seems beneficent, a limitless supply of energy that Whitehall civil servants eagerly grasp at, is actually deadly. As an alien Axon creature explains, "Axonite can absorb, convert, transmit and program all forms of energy." It sucks a human being into a dry husk and, if left unchecked, would do the same to the whole planet, as the Axon creature proclaims, "Slowly we will consume every particle of energy, every last cell of living matter. Earth will be sucked dry!"

The Pertwee era of *Doctor Who* also imbued energy with a distinctive and vital narrative function, placing it as central to an emerging backstory and mythology of the Doctor's own race of people the Time Lords. The question of what powered the TARDIS was, as noted above, raised in the program's very first episode in 1963, and in the second serial mercury was also explained as intrinsic to the ship's functioning. "The Three Doctors" (1972–73) defines energy as central to the technology that enables time travel itself. Viewers learn that the "secret of time travel" was a Time Lord discovery but "in order to make it a reality we had to have a colossal source of energy," which came from engineering power from a black hole. In "The Three Doctors" as in "The Claws of Axos" the same notion of energy as something that takes rather than gives recurred, and the narrative of this serial, in that three incarnations of the Doctor are required to join forces, is because of the loss of "vital cosmic energy" that is vanishing through a black hole.

From person to person, energy transfer was also exploitative. "The Savages" (1966) delineates two contrasting societies, a technological urban civilization of the Elders and the cave dwellers derided as "savages." The high technology and urban pleasures of the society depend on an ongoing energy transfer process, in which the intelligence and creativity of the so-called savages is drained off to enhance the Elders and their entire society. A similar principle of energy transfer provided the plotting in "The Krotons" (1968–69), referring to a form of "mental energy" that the Krotons could take from intelligent young people.

In other scripts in the same decade, the notion that a source of energy was more like a parasite taking away than a gift giving recurred. A large but uninhabited city in "Death to the Daleks" (1974) is a parasite, having drained energy from an entire world. The TARDIS, as well the spaceship and weaponry of Daleks and human Marine Space Corps members all fail as the city sucks their energy away. The serial "The Talons of Weng-Chiang" (1977) evocatively describes dead bodies as "like old leaves," after a process of organic distillation has taken energy from their bodies and transferred it to another. In "Image of the Fendahl" (1977), as in "The Claws of Axos," characters find a human body drained of not only life but also all forms of energy and any binding agent that would hold tissues together. All the binding force has gone and all that remains is a husk. In "The Horns of Nimon" (1979–80) the bull-like, but parasitic, Nimon drain entire worlds of life and also human bodies of their vitality. The Doctor's companion Romana theorizes that, "the Nimon feeds by ingesting the binding energy of organic compounds such as flesh." This story shows a complex interplay of energy as both created and taken. The Nimon use hymetusite, described on screen as an immensely powerful source of energy, to artificially engineer a black hole and jump from world to world. By doing so, they consume energy parasitically. "The Horns of Nimon" came soon after other serials where parasitic energy users destroyed entire worlds. "The Pirate Planet" describes how a marauding bionic pirate captain can "ransack entire planets" of their energy to feed time dams.

Energy in Actuality

Doctor Who brought different sources and types of energy onto the screen, often coupled with avaricious or misguided civil servants and politicians. Energy could be characterized using conventional terminology (nuclear) or fictional (taranium) and frequently as destructive and exploitable. In some serials, those anxious to locate energy and who unwisely planned its use were recognizable twentieth century bureaucrats, but the type could modulate across settings. The renaissance nobleman Count Federico in "The Masque of Mandragora" (1976) and the general Soldeed in "The Horns of Nimon" were unwise exploiters of energy. While scientists in *Doctor Who* tampered with energy, sought and unleashed new sources, or were threatened by them, energy in the real world was part of a social, political and scientific context leading from the 1960s into the 1970s. A complex interplay of factors, including the Yom Kippur War and the subsequent punishment of Israel's allies by the OPEC countries by cutting oil supplies, limited energy supplies not just in England but also elsewhere in Europe.

Conversely, energy in abundance rather than its absence was also a source of social anxiety, with the nuclear arms race escalating as first the United States followed by the Soviet Union and the United Kingdom gained nuclear capability. Not only bombs but also other uses of nuclear technology registered in public fears. In 1957, the Windscale Nuclear facility in Britain caught fire, prompting wide scale concerns about the release of radioactive material into the environment (Arnold 1992). However, atomic developments were also couched in a positive frame of reference, and anxiety and optimism could even occupy the same space. Science fiction cinema in the 1950s could narrativize nuclear disaster, but between the A and B films, the commercials shown in cinemas could espouse the benefits of nuclear technology. Nuclear expertise also occupied an optimistic space in public discourse as a field of development that would restore the economy that the Second World War had ravaged (Jones 2017).

Politically, the discourse of nuclear optimism also aligned with the reforming Labour governments. When he was Prime Minister in the 1960s, Harold Wilson articulated a vision of a modern post-war Britain forged in the "white heat" of a scientific crucible (Hughes 2015: 49). To turn that into a reality, government measures included the creation of a Ministry of Technology, headed by Anthony Wedgwood Benn, and the investment in nuclear research, while also attempting to stop a brain drain of scientists to the United States (*Birmingham Post* 16 November 1967, p.1). The government promoted Britain throughout the Commonwealth as undergoing "technological revolution" (*Birmingham Post* 27 February 1967, p.9). The government activity also registered in popular culture, especially science fiction on 1960s British television. Benn's Ministry of Technology had fictional analogs. In the outlandish adventure series *The Avengers* (1961–69), the Ministry of Technology Cybernetic & Computer Division, and the Ministry of Technology's Neoterric Research Unit, were sites of unsettling and uncontrollable technological threats.

However, asserting a technological revolution was frequently undercut and in tension with anxiety about a brain drain, in that the British government was not investing enough into technology, and British research in state of the art fields such as lasers was falling behind the United States (*Birmingham Post* 8 September 1969, p.7). Wilson's intentions were policies that followed in the wake of C.P. Snow's notion of the "two cultures" (expressed in a speech of that title in 1959) within British intellectual life, and in particular Snow's criticism that the humanities were unduly privileged. Consequently, argued Snow, British political elites were not coping with the scientific revolution, when science had contributed to the allied victory against Germany. Snow's ideas followed in the wake of astronomer Fred Hoyle's suggestion for scientists to supplant politicians as being responsible for running the country (Parker et al. 1999). Wilson had intended his "white heat" speech to set an agenda for modernization, not simply through scientific development but in attitudes of political elites and society in general to science itself. However, the words carried a dangerous if unintended double meaning; after all, what else would the detonation of nuclear bombs during the Cold War have been except a lot of white heat as hydrogen bombs detonated? The poet Cecil Day Lewis articulated fear that a nuclear bombardment could obliterate the entire country, considering it the "overriding issue of our time" (cited in McLoughlin 2018: 5). The Bay of Pigs and the Cuban Missile Crisis in the early 1960s served to reinforce Britain's preoccupation with nuclear oblivion, as did occasional accidents such as the spillage of radioactive materials on the A5 in 1967 (*Birmingham Post* 19 August 1967, p.1).

Energy, or its absence, was of sufficient concern by the early 1970s that the then Prime Minister Edward Heath created an entire government department just for energy and appointed Lord Carrington as the Secretary of State for Energy (*Birmingham Post* 9 January 1974, p.1). Carrington therefore became responsible for coal, oil, electricity, and atomic power, and especially for making North Sea oil flow (Campbell 2013). Accounts of Edward Heath's last weeks as prime minister note his overwhelming preoccupation with the oil, even to the exclusion of other crises (Beckett 2009). The Time Lords watched with dismay in "The Three Doctors" as their energy drained away, owing to the hostile actions of the renegade Omega; British politicians in the same year watched in horror as Middle Eastern governments cut oil supplies (*Birmingham Post* 14 December 1973, p.9). At the same time, three-day weeks came into effect, as electricity supplies dwindled as miners went on strike, meaning there was insufficient coal for power station furnaces (*Birmingham Post* 14 December 1973, p.9; *Financial Post* 14 January 1974, p.1). Up to two thirds of the workforce was affected by the fuel and power shortages and unable to work a full week by February 1974 (*Illustrated London News* February 1974). Concomitants of the three-day week included rising unemployment and social disorder (Venn 2016).

By 1974, a full-blown energy crisis occupied Edward Heath's government, which anxiously sought ways to achieve savings in electricity consumption and fretted about fuel and energy (*Birmingham Post* 10 January 1974, p.2). The Heath administration spoke of the country as starved of energy (Holmes 1997). The notion of an energy crisis entered common parlance as did concern over the long-term resources needed for Britain's energy needs. The bureaucratization of energy, in government departments, ministry officials and politicians, features in *The Avengers*, where a generic ministry associated with power and technology oversaw energy research. Significantly, the ministry came across as misguided or haphazard. "The Positive Negative Man" (a 1967 *Avengers* episode) suggested the government department had discontinued funding into "broadcast power," an energy innovation that the series showed as actually working. The James Bond film *The Man with the Golden Gun* opened in 1974 with the energy crisis as the main thrust of its narrative. Bond was in pursuit of a villain who had found a way to harness solar energy, while M, Bond's superior, spoke anxiously of diminishing supplies of oil and coal and demands placed on the British tax payer (Upton 2014).

What is notable is that for some commentators, Britain's energy crisis was a political and industrial situation that could be interpreted using themes and scenarios from science fiction. In as much as contemporary science fiction may have gained inspiration from Britain's energy woes and Whitehall's efforts to find limitless available power, the inspiration flowed in the opposite direction too. For example the newspaper columnist Alexander McDonald, writing after the Yom Kippur war when Middle Eastern states were shutting off the oil supply to the west, turned to classic science fiction to explain the waves of fear rippling through Whitehall: "In the pre-war film of H.G. Wells' *Things to Come* there was a scene showing a world without oil. Civilisation had collapsed following a cataclysmic world war and horses were being used to pull cars around" (*Birmingham Daily Post* 10 December 1973, p.6). In his opinion, the science fiction film (made in 1936) would now make uncomfortable viewing for the ministers and civil servants who were afflicted by a "distinct touch of Doomsday panic."

The energy crisis also inspired television science fiction. *Doctor Who* serials made in 1973 reflected the concerns surrounding the energy crisis. The Wholeweal community in "The Green Death" is a body of scientists looking for alternative and clean sources of energy that are not oil or coal. The next year "Invasion of the Dinosaurs" showed an idealistic scientific community who believe they have left the planet Earth to seek a new and clean world, because of the damage caused by industrial pollution. The themes in these serials cohere with other television outputs from the period. "Lighthouse Keeping Loonies," a 1975 episode of *The Goodies* (1970–82), noted and drew comedic science from Britain's problems with getting oil from the Middle East when they find a large oil supply underneath a remote lighthouse, which ignites and shoots the lighthouse into space. One of the Goodies describes the discovery of the oil in terms alluding to the actions of the OPEC nations and resentment of Middle Eastern governments, and exclaims with racist idiom, "that's one in the eye for the greasy Arab!"

Here the comedy and science mingle. A darker tone prevails in Nigel Kneale's *Quatermass* (1979), a serial reviving his 1950s science fiction adventures, set in the future but ineluctably a product of 1970s energy concerns. As James Chapman (2006) points out, its foregrounding of social and urban collapse, because of the loss of fuel, mirrors contemporary concerns that registered in popular culture. At one point Professor Quatermass remarks of the social disorder that the North Sea oil was going to "put everything right," but the series shows even that energy source failed to bring civilization back from the brink.

Science fiction could inspire descriptions of energy crises, and science fiction took inspiration from government departments and government concerns that the power would go out. Although *Doctor Who* is a patchwork of different editorial decisions, writers' imaginations and production activity, many stories consider how energy would simply destroy the world. As Derek Johnson has indicated, science fiction content was presented in early BBC television broadcasts from the 1930s, a point of origin significantly earlier than the 1950s Quatermass serials. By the 1950s, science fiction was routinely on British television, but the accent was often prophetic or troubling. "Underground" (1958), an installment of the *Armchair Theatre* television anthology series (1956–74), dramatized a group of survivors living in the London Underground after a nuclear war on the surface. *The Avengers* presented the possible threat of nuclear annihilation as part of dramatic scenarios too. Protagonists Steed and Dr. Gale, Mrs. Peel or Tara King visited underground nuclear shelters and investigated threats to Britain's early warning radar systems and the security of missile bases (Chapman 2002). In one episode, when the radar system erroneously indicated a missile strike was imminent and retaliation was only narrowly aborted, Steed commented to Gale that they "would've been mutating by now" if the bombs had dropped ("Dressed to Kill," 1963).

The BBC intended to dramatize (in mock documentary format) the effects of thermonuclear bombardment on Britain in *The War Game* (released in cinemas 1966), although they ultimately delayed the transmission originally scheduled for 1965 because of political concerns (Piette 2009). In 1963–64, *Doctor Who* portrayed the implications of neutronic war in "The Daleks," the first serial to portray the Daleks and their antagonists the Thals, both (different) outcomes of genetic mutation caused by radioactive poisoning from the neutron bomb. Bunce (2010) points to the significance of the terminology, as the notion of a neutron bomb that would destroy life not matter and emit high doses of radiation had featured in the press since 1959. Early in the serial, the Doctor and his companions begin to fall alarmingly unwell, and exploring an abandoned city, they hear an ominous ticking noise and see a Geiger counter. The Doctor realizes how "that explains a lot of things, doesn't it. A jungle turned to stone, the barren soil and the fact that we're not feeling well" as "the atmosphere here is polluted with a very high level of fallout, and we've been walking around in it completely unprotected." References elsewhere in the story to radiation sickness and mutation contribute to the serial's ominous presentation of the aftermath of nuclear war. Some subsequent serials showed the impact of atomic war. The Doctor averts the use of atomic weapons at the Battle of Hastings in 1066 in "The Time Meddler" (1965). In "The Faceless Ones" (1967) an alien species, the chameleons, have lost the molecular cohesion of their bodies after an atomic explosion.

The irradiated surface of the planet was an alarming vision in science fiction that intersected with public discourse. The British reading public learnt from the press that the Swedish people and government were building a vast underground cave network to survive an atomic war (*Coventry Evening Telegraph* 21 June 1966, p.11). The British government sought to reassure its own people that nuclear power stations were not a threat to human existence like the atom bomb was (*Coventry Evening Telegraph* 19 October 1967, p.16) and to ensure that anarchy would not break out if there was an atomic attack on the British Isles (*Birmingham Post* 26 July 1967, p.2).

These examples return attention to *Doctor Who*'s preoccupations with energy. The program's own creative crucible is marked by contrasting political and social fears about science. Early serials in which the TARDIS landed on a dead world (in "The Daleks") or in a London shattered by bombs (in "The Dalek Invasion of Earth") spoke to recent wars and fears of yet more destructive atomic conflict. In these early serials, there was too much energy, but later serials also showed Whitehall mandarins concerned that there was too little. Energy that destroyed, that was parasitic, or came from sources so efficient that humanity would be superseded recur in narratives both earth-bound and alien. These stories speak to developments, social, scientific and political, in which science fiction reflected the preoccupation that energy was needed but unsettling.

REFERENCES

Newspapers

Birmingham Post
Coventry Evening Telegraph
Financial Post
Illustrated London News

Other Sources

Arnold, L. (1992) *Windscale 1957: Anatomy of a Nuclear Accident*. London: Palgrave Macmillan.
Beckett, A. (2009) *When the Lights Went Out: What Really Happened to Britain in the Seventies*. London: Faber & Faber.
Bunce, R. (2010) The *Evil* of the Daleks. In: Lewis, C., and Smithka, P. (eds.) *Doctor Who and Philosophy: Bigger on the Inside*. Chicago and La Salle: Open Court, pp. 339–350.
Campbell, J. (2013) *Edward Heath: A Biography*. London: Random House.
Chapman, J. (2002) *Saints and Avengers: British Adventure Series of the 1960s*. London: I.B. Tauris.
Chapman, J. (2006) *Quatermass and the Origins of British Television SF*. In: Cook, J.R., and Wright, P. (eds.) *British Science Fiction Television: A Hitchhiker's Guide*. London: I.B.Tauris, pp. 21–51.
Harmes, M. (2014) *Doctor Who and the Art of Adaptation*. Lanham, Maryland: Rowman and Littlefield.
Holmes, M. (1997) *The Failure of the Heath Government*, 2nd edition. London: Macmillan.

Hughes, G. (2015) *Harold Wilson's Cold War: The Labour Government and East-West Politics, 1964–1970*. London: Boydell and Brewer.
Jones, M. (2017) *Science Fiction Cinema and 1950s Britain: Recontextualizing Cultural Anxiety*. London: Bloomsbury.
McLoughlin, K. (2018) Introduction. In: McLoughlin, K. (ed.) *British Literature in Transition, 1960–1980: Flower Power*. Cambridge University Press, pp. 1–30.
Norris, R.S., and Kristensen, H.M. (2012). The Cuban Missile Crisis: A nuclear order of battle, October and November 1962. *Bulletin of the Atomic Scientists* 68(6): 85–91.
Parker, M., Higgins, M., Lightfoot, G., and Smith, W. (1999) Amazing Tales: Organization Studies as Science Fiction. *Organization* 6(4): 579–590.
Piette, A. (2009) *Literary Cold War, 1945 to Vietnam*. Edinburgh: Edinburgh University Press.
Upton, B. (2014) *Hollywood and the End of the Cold War: Signs of Cinematic Change*. Lanham, Maryland: Rowman and Littlefield.
Venn, F. (2016) *The Oil Crisis*. London: Routledge.
Warrick, P.S. (1980) *The Cybernetic Imagination in Science Fiction*. Cambridge, Massachusetts: MIT Press.

Translation by TARDIS

Exploring the Science Behind Multilingual Communication in Doctor Who

MARK HALLEY *and* LYNNE BOWKER

Introduction

The ability to communicate easily between multiple languages has long been a staple of science fiction. The "Babel fish" in Douglas Adams' *Hitchhiker's Guide to the Galaxy* (1978–80), the "universal translator" of *Star Trek* (1966–69), and of course, the "translation circuit" (occasionally called the "translation matrix") in *Doctor Who* are well-known examples of devices or organisms used to circumvent the difficulty of translating between alien languages (unless of course, that problem is essential to the plot). But how does science fiction translation stack up to *bona fide* machine translation? The challenge of linguistic diversity—even here on Earth—has long inspired linguists and philosophers, and more recently mathematicians, engineers, and computer scientists, to search for theories and devices that could address problems resulting from the multiplicity of languages. As early as 1933, two patents were filed—one by Georges Artsrouni in France, and the other by Petr Trojanksij in Russia—for electromechanical devices that could act as multilingual translation dictionaries (Hutchins 2004). Neither received full development, however, and it was not until the introduction of digital computers after World War II that the research program in machine translation produced actual tools.

Machine translation has since undergone several paradigm shifts, but has it grown to resemble so-called science fiction technology more closely, or has it developed in another direction? This essay explores the evolution of machine translation from the 1940s to the present day to see how machine translation technology compares to what we know about *Doctor*

Who's translation circuit. More specifically, we look at what machine translation and the translation circuit have in common, where they differ, and how the characterization of translation technologies in *Doctor Who* affects the public perception of actual machine translation. In this short essay, we cannot consider all the occurrences of the translation circuit in *Doctor Who*. Rather, we focus on presenting key features of machine translation technology and then use illustrative examples of the translation circuit that demonstrate similarities with or differences from machine translation.

But let us first begin with an overview of how translation came to be in *Doctor Who*. In the first episode ("An Unearthly Child," 1963), issues of interplanetary language and communication seem to be a non-issue. The characters, despite not all being humans or even earthlings, communicate without issue and with no indication to viewers—or the characters—that translation of any sort is taking place. The episodes that follow introduce creatures from different times and worlds, but communication still occurs without complication. Viewers were left to speculate on how it was that ancient cave people used the same spoken language as time travelers from Gallifrey and high school teachers in London, not to mention how evil beings from the planet Skaro could read a note written by Susan, the Doctor's granddaughter ("The Daleks," 1963–64). However, as the series continued, the writers began to indicate that some form of translation was—or was not—at play. This is made most evident by the apparent inconsistencies of the translation circuit. When the Doctor travels back to the French Revolution in "The Reign of Terror" (1964), he communicates in French with eighteenth-century Parisians with relative ease, although later in the serial his companions Ian and Barbara seem to attempt speaking in a French accent and are aware of its difference from English. However, when he finds himself in World War I in "The War Games" (1969), the Doctor is unable to understand French, despite demonstrating an ability to speak German and having even spoken some French himself in "The Moonbase" (1967). As recently as "The Haunting of Villa Diodati" (2020), we see a character speaking French, despite the TARDIS—and presumably the translation circuit—being nearby. Just what is going on, and how do the writers explain the complexity of intergalactic communication in *Doctor Who*?

To start, consider the first explanation of interlingual communication, which did not occur until "The Masque of Mandragora" (1976). Up until that point, viewers were left to their imagination when considering how Time Lords, Daleks, humans of different nationalities, and all other manner of living species from across the universe were (sometimes) able to communicate with relative ease. When the Doctor's companion, Sarah, suddenly asks how it is that she can understand Italian, he realizes that her mind had been possessed. As he explained, "Well, I've taken you to some strange places before and you've never asked how you understood the local language. It's a Time Lord's gift I allow you to share. But tonight when you asked me how you understood Italian, I realized your mind had been taken over."

Throughout the series, we gradually learn more about translation in the Doctor's universe. For example, the Doctor notices a Somersetshire accent in "The Ribos Operation" (1978) and a French accent in "The Girl in the Fireplace" (2006). Recall too when the Doctor meets Vincent van Gogh, who astutely asks, "That accent of yours. You from Holland like me?" Another apparent imperfection of the translation circuit arises in "Four to Doomsday" (1982) when Tegan, a white Australian, implausibly speaks the Indigenous Tiwi language and proceeds to interpret between the Doctor and Kurkutji, an Aboriginal man. As Orthia (2013) noted, it is unlikely that Tegan would have even known which one of the hundreds of Indigenous languages Kurktuji spoke, let alone have been able to speak it fluently. This situation further begs the question as to why the translation circuit does not enable the Doctor to communicate with Kurkutji directly.

Throughout *Doctor Who*, there are numerous moments in which we are left to question the apparent science behind the translation circuit. Although it sometimes works so well that we—and the Doctor's companions—do not even question what is happening, there are other times in which it seems to stop working altogether, or to work inconsistently or in unexplained ways. Why do we hear Liz Shaw speaking French in "The Ambassadors of Death" (1970)? Does the translation circuit not work for the Chinese Hokkien language, as the Brigadier is unable to understand the Doctor, who speaks the language with ease in "The Mind of Evil" (1971), or was the Doctor's use of a foreign language intentional? Just a few years later (for us earthlings), the Doctor speaks Mandarin in "The Talons of Weng-Chiang" (1977),[1] suggesting that he is aware of when the translation circuit is and is not capable of meeting his translation needs. Inconsistencies occur across writers, incarnations of the Doctor, and series, and occasionally even within a single episode. For instance, in "The Two Doctors" (1985), we hear the Doña Arana speaking Spanish, and she remarks upon the ability of Shockeye the Androgum to speak English. However, in the same story we learn from the Doctor that the Androgums and Sontarans do not speak Spanish. Why is the translation circuit unable to overcome this linguistic barrier?

In the rest of the essay, we explore the history and science of machine translation in our—to the best of our knowledge—TARDIS- and Time Lord–free world. In doing so, we also draw parallels and distinctions between machine translation and translation offered by the translation circuit.

A Tool by Any Other Name? The Terminology of Translation

Before describing machine translation, we will explain a terminological issue. Language professionals clearly distinguish between translation, which refers to transferring the content of a *written* text from one language to another, and interpretation, which describes the transfer of content from one language to another in an *oral* or *signed* mode (Munday 2016). Indeed, Pöchhacker (2004: 10) describes interpretation as a special type of translation, noting that "interpreting can be distinguished from other types of translational activity most succinctly by its immediacy: in principle, interpreting is performed 'here and now' for the benefit of people who want to engage in communication across barriers of language and culture." In *Doctor Who*, the so-called *translation* circuit mostly provides immediate mediation for oral communication, and less frequently for written communication, and so, it might be more appropriately termed an *interpretation* circuit.

Although machine translation has long focused on written text, this technology has recently been combined with speech recognition and speech generation tools (e.g., Dalvi et al. 2018). Taken together, a cascaded speech recognition-to-machine translation-to-speech generation model could simulate an automatic interpretation system, which more closely resembles how the translation circuit functions. Meanwhile, machine translation can combine with image processing technology (e.g., a camera and optical character recognition tool) to translate signs or handwritten notes, such as the addresses on envelopes (Wen 2019). This is similar to the way the translation circuit approaches the translation of written material, such as in "The Fires of Pompeii" (2008), when the Doctor takes his new companion Donna to ancient Italy. At first, Donna is skeptical that they are actually there because she sees an English-language advertisement. The Doctor explains that it is the TARDIS translation circuit that is making the sign appear as if it were written in English, as well as converting the speech of the locals. The translation circuit also experiences a "lag" when faced with written text, such as a prayer leaf that contains the name "Melody Pond," which takes some time to be translated as "River Song" by the translation circuit ("A Good Man Goes to War," 2011). This type of lag would be true of machine translation combined with image processing too, since multiple technologies must work together such that the output of one becomes the input to the next. However, in "Orphan 55" (2020), viewers see untranslated Russian writing, leaving us to question whether the translation circuit is working its linguistic magic on its own time or is simply unable for some reason to translate written Russian.

We noted previously that the translation circuit is inconsistent when providing interpretation, and it is similarly inconsistent when translating text. For example in "Terror of the Zygons" (1975), the Time Lord gift does not help Sarah Jane when she attempts to read Medieval Latin records. In contrast, in "State of Decay" (1980), the translation circuit is able to translate ancient texts, although the Doctor then applies the Law of Consonantal Shift to the names of feudal lords and determines that they are not who they seem to be (e.g., Lord "Sharkey" becomes "Zargo").

Breaking Down Language Barriers or Breaking Down Borders? The Sociocultural Context of Machine Translation

When we think about translation, it is often in a feel-good way, where translation breaks down language barriers and allows people—or aliens—to communicate for largely positive reasons. Baker (2005: 9), for example, describes this as translation's "master narrative":

> In translation studies today, we have a master narrative of the translator as an honest intermediary, with translation repeatedly portrayed as a force for good, a means of enabling dialogue to take place between different cultures and therefore (the logic goes) improving the ability of members of these different cultures to understand each other.
>
> Thus, communication, dialogue, understanding, and indeed knowledge are assumed to be "good" in a moral sense. They lead—unproblematically—to justice, peace, tolerance, progress.

To this end, the translation circuit is repeatedly described as "a Time Lords' gift" (e.g., in "The Masque of Mandragora") or a "gift of the TARDIS" (e.g., in "The End of the World," 2005), suggesting perhaps that it is intended to be used primarily for positive ends and thus perpetuating the "master narrative" described by Baker. One is left to wonder whether, without the translation circuit, the Doctor would be able to so adeptly save the day repeatedly. However, Baker (2005) goes on to argue that this narrative obscures some of the real issues that may arise, such as a deliberate will to misunderstand, and the complex role that translators play in situations of conflict, among others. *Doctor Who*, meanwhile, remains largely faithful to the master narrative, even though this is not always how translation, including machine translation, works in the real world.

With regard to machine translation's origins, early work appeared to be well motivated on the surface. Warren Weaver, an American mathematician and Rockefeller Foundation employee, is credited with launching machine translation research through the release of "Weaver's

Memorandum" (1949). In it Weaver laments the "multiplicity of language" that "impedes cultural interchange between the peoples of the earth, and is a serious deterrent to international understanding," and he suggests that electronic computers could be used to address the resulting "world-wide translation problem." Weaver's motives in suggesting machine translation appear noble; however, cryptanalysis and code-breaking techniques developed during World War II partly inspired his recommendations, and early funding for machine translation came mainly from the US Defense Advanced Research Projects Agency (DARPA), the Central Intelligence Agency (CIA), and other military sources. Moreover, the strategic choice of languages (e.g., Russian-to-English for the Georgetown IBM experiment in the 1950s) is telling, reflecting the Cold War anxieties of the time (Hutchins 2000).

However, machine translation proved more difficult to crack than researchers had anticipated. As progress stalled, research sponsors formed the Automatic Language Processing Advisory Committee (ALPAC) to evaluate the progress of machine translation research. This Committee released the ALPAC Report (1966), which unfortunately sounded a veritable death knell for machine translation. Key findings included the assessment that machine translation output was expensive, of poor quality, and too slow, as well as the observation that there were enough professional translators to meet market needs. The Committee recommended that it was no longer worth investing in machine translation research and that it would be preferable to focus on developing tools to *assist* rather than replace translators. The Committee, it would seem, did not have science-fiction-worthy aspirations, and following the ALPAC Report, funding for machine translation research virtually dried up.

Gradually, however, the world changed, and interest in machine translation research rekindled toward the end of the twentieth century, inspired more by commercial interests than military ones. The Internet was significant because it facilitated international e-commerce, which resulted in a new area of translation known as "localization," where products (e.g., software packages, websites) are linguistically and culturally adapted to enable customers to interact with them in their own language (Esselink 2000). With the Internet and globalization, the volume of texts to be translated increased exponentially, catapulting machine translation from being primarily a research-driven area to being a needs-driven one.

Although the precise origins of the translation circuit in *Doctor Who* are unclear, its use does not seem to have been motivated by military or commercial interests. Further, we note that Time Lords generally use the TARDIS itself for observation and non-intervention (with notable exceptions from the Doctor, the Master, the Meddling Monk, and the Rani). The translation circuit is used altruistically to facilitate communication and even to avert crises. For example, recall when the Doctor found himself in the twenty-fourth century on a spaceship hurtling toward earth, under threat of being shot down by the Indian Space Agency in "Dinosaurs on a Spaceship" (2012). Arriving on the ship with his usual companions, as well as a game hunter, dinosaurs, and Queen Nefertiti from Ancient Egypt, he adeptly diverts the missiles toward the ship belonging to the episode's villain. Would the Doctor and his companions have been able to save the dinosaurs without the help of the translation circuit? It is doubtful that the Doctor's twenty-first century British companions spoke Nefertiti's fourteenth century BCE Egyptian. In a way, it is the ubiquitous nature of the translation circuit that allows the Doctor to accomplish most of their feats. At present, machine translation is not as ubiquitous as the translation circuit, but it is certainly increasing its reach. For instance, if you use Google Chrome as your browser, it will automatically offer to translate web pages into your preferred language, and so will social media applications such as Facebook and Twitter, which have embedded machine translation widgets.

Moreover, we see that in the twenty-first century, machine translation is beginning to make inroads into the humanitarian arena too. Owing largely to the availability of free, online systems, machine translation is used in a wider range of situations, including applications for social justice and humanitarian efforts. For instance, researchers are investigating how effectively machine translation can be deployed as part of a crisis response, such as to facilitate communication following natural disasters like earthquakes or during health-related crises like Ebola (e.g., Federici and Al Sharou 2018; O'Brien and Cadwell 2017). So while machine translation may not yet be up to the task of saving the universe, it is certainly gaining ground.

How Does Machine Translation Work? From Rules and Statistics to Artificial Intelligence

Dialogue in *Doctor Who* describes the TARDIS's translation circuit as a "gift" and its precise origins are unclear. It appears to function automatically when needed, though there are exceptions when it does not work properly or at all. For example, when the Doctor and his companions encounter a deaf British Sign Language user, they find that British Sign Language had been "deleted" from the translation circuit's memory, only to be replaced by semaphore, a communication system that uses flags to share messages ("Under the Lake," 2015).² In contrast to the effortless automaticity of the translation circuit, machine translation is the result of decades of painstaking labor during which developers used different approaches to tackle the challenge of machine translation. Early approaches are characterized as "rule-based" because much energy was spent building large bilingual dictionaries and formalizing sets of grammatical rules that detailed how lexical items could be combined into sentences. These systems used dictionary look-up to provide word-for-word translations and then employed some local reordering rules to fix a limited number of errors (Poibeau 2017). However, a word-for-word translation rarely works. Consider the American Sign Language idiom "FISH SWALLOW," which is used to suggest that another person is gullible. One can imagine the confusion that might ensue if the translation circuit or a machine translation system were to offer a word-for-word translation rather than a more idiomatic or meaning-oriented translation.

Second-generation machine translation systems adopted a more sophisticated rule-based transfer approach where the original language sentences were analyzed into an intermediate representation, which was then used to generate translated sentences (Poibeau 2017). While the results were more promising, a different system had to be constructed for each language pair. Hoping for a system that could work for virtually any language pair—as the TARDIS's translation circuit appears to nearly accomplish—researchers next attempted to develop interlingual systems, which sought to employ a so-called universal language. In interlingual systems, the representation of the content of the source-language sentence would ideally no longer bear traces of the source language, and it could be used to generate a sentence in any target language (Kenny 2018). In other words, a true interlingua would be natural-language independent and would involve some kind of universal notation, capable of conveying any meaning expressible in any source or target language covered by the system. Indeed, an interlingual system would seem to be approaching the capabilities of the TARDIS translation circuit; however, the approach has never been deployed on a large scale, and researchers ultimately found it impractical and even theoretically misguided. It therefore remains in the realm of science fiction, at least for the time being!

While the Doctor and his companions chatted away with relative ease thanks to the translation circuit, the rest of us limped along with rule-based machine translation systems, with their inherent challenges and limited successes, for approximately half a century. However, in the new millennium, corpus-based approaches were introduced for which no linguistic knowledge had to be encoded. Rather, the main requirement is to provide an enormous volume of previously translated texts in the form of parallel corpora. It is therefore relatively easy to extend corpus-based machine translation systems to work with new language pairs and it brings us a step closer to the TARDIS's translation circuit, which seems to have few problems taking on new languages at the drop of a hat (with some exceptions, such as the ancient language used in "The Impossible Planet/The Satan Pit," 2006).

The main corpus-based approach that dominated machine translation research from the mid–2000s was statistical machine translation, where systems are trained using parallel corpora and make substantial use of probability calculations (Koehn 2010). While statistical machine translation systems typically produce higher quality output than rule-based machine translation systems, they nonetheless have weaknesses. Sometimes, like the translation circuit, they translate a given word inconsistently or simply fail to translate it at all. They also have difficulty translating polysynthetic languages (e.g., Tiwi) or highly inflected languages (e.g., Latin) because they have no principled way of handling grammatical agreement. We saw previously that the translation circuit could not translate Tiwi in "Four to Doomsday," and that it failed to help Sarah Jane read Medieval Latin records in "Terror of the Zygons." We can therefore identify similarities in the difficulties faced by both the translation circuit and statistical machine translation systems.

In 2016, an approach known as neural machine translation emerged, which uses artificial intelligence and machine learning techniques (Way 2019). Like statistical machine translation systems, neural machine translation systems learn to translate from parallel corpora; however, they use very different computational methods to do so. An artificial neural network is a sort of information processing system that is inspired by the way biological nervous systems, such as the brain, process information. A main difference between neural machine translation and statistical machine translation is that when researchers present training material to the deep learning algorithms in an artificial neural network, they do not necessarily tell them what to look for. Instead, neural machine translation systems find patterns themselves (e.g., contextual clues around the source sentence).

Of all the approaches, neural machine translation most closely resembles the translation circuit in *Doctor Who*. Each is something of a black box, and while neural machine translation uses an *artificial* neural network, this is inspired by a biological nervous system. Although Time Lords are not humans and they have some physiological differences from humans (e.g., two hearts, lower core body temperature, telepathic abilities), they are presumably still biological entities with some kind of brain-like organ to which the translation circuit connects.

Buyer Beware: Comparing Uses of the Translation Circuit and Machine Translation

Now that we have explored the sociocultural context in which machine translation was developed, as well as different underlying approaches to machine translation, and compared these to what we know about the translation circuit, let's compare the use of the two at a more fine-grained level. Although *Doctor Who* does not provide many details about the origins or workings of the translation circuit, a few interesting tidbits are shared that we can compare to what we know about machine translation.

One notable thing about the translation circuit is that while it appears to function flawlessly most of the time, there are occasions where it does not work. Several reasons are given for this. For instance, the translation circuit does not appear to work when the Doctor is debilitated ("The Christmas Invasion," 2005), when the TARDIS is sabotaged ("The Time Monster," 1972), or when the speaker is out of telepathic range as in the *Doctor Who* novel *The Eye of the Giant* (Bulis, 1996). Further, the translation circuit (and TARDIS more generally) is inconsistent and perhaps even unreliable (e.g., the TARDIS controls are not consistently isomorphic). Since machine translation systems are external to their users, the comparison is rather superficial, but we could nonetheless say that machine translation systems will not work if users are out of range (e.g., away from the keyboard, mouse, screen, or microphone). Similarly, if users cannot operate their device, they will not be able to use the machine translation system. Finally, if the computer hardware or software is damaged, the machine translation system will not be accessible.

The Doctor is sometimes confronted with languages that are ancient, dialectal, or non-verbal. An example of the latter is the "eyebrow" language in "Spearhead from Space" (1970), which the translation circuit cannot process. As we have learned, machine translation systems can only process languages for which they have resources available (e.g., lexicons, rules, corpora). They cannot tackle languages for which they have no resources. This would sometimes seem to apply to the translation circuit also. Recall the earlier example in which British Sign Language had been deleted from the translation circuit's memory store. Without the resources to process the language, the translation circuit is of no use.

More interesting than these superficial observations, however, are cases where the translation circuit could provide only a partial translation, or where only the essence of the message was transferred. For example, in Christopher Bulis' *Imperial Moon* (2000), the Doctor provides a particularly complex explanation about time travel to his companion. As the Doctor puts it, "We will have been here before—I'm afraid the English language doesn't really have the correct tenses for time travel." This is much closer to the reality of machine translation and could be compared to what is known as "gisting" in machine translation circles: the quality of the machine translation output is not polished, but it may be good enough for readers to get the gist of the original text's message. In rule-based machine translation systems, problems would occur if the bilingual dictionary had gaps, or if the grammar rules were incomplete or improperly specified. In such cases, the machine translation output would likely be of poor quality. For corpus-based approaches, the bigger the training corpus, the more likely it is that the machine translation system will propose a reasonable translation. In the case of low resource languages (i.e., languages for which there is little training data available), the quality of the results may be low as well, but the system will at least try. The only time a machine translation system will fail to produce anything is if it is not given the needed resources (e.g., lexicon and grammar for rule-based machine translation, training corpora for statistical or neural machine translation). To go beyond gisting, however, and to get a professional quality translation, the raw machine translation output often needs to be edited by a human. However, if the goal is simply to get the "gist" of the original text, rather than to publish the translation, then the raw machine translation output may suffice. As the *Imperial Moon* example demonstrates, the translation circuit must also resort to gisting at times.

Regardless of the underlying approach to machine translation that is employed (i.e., rule-based, statistical, or neural), machine translation systems typically produce better results when the input text is "controlled" to make it more translation-friendly. In machine translation, a controlled language has a limited lexicon and a restricted syntax. Sometimes this occurs naturally, such as weather forecasts, which draw on a naturally restricted range of vocabulary and grammatical structures. The METEO system was a highly successful machine translation system that translated weather bulletins for Canada's ministry of the environment (Langlais et al. 2005). While weather bulletins represent a spontaneously occurring sublanguage, it is possible to impose restrictions by design, creating a controlled language. An example is Caterpillar Technical English, a controlled language used to write the technical documentation for Caterpillar's products. Controlled languages are easier for machine translation systems to translate because they contain fewer lexical or syntactic ambiguities, and therefore they offer fewer opportunities for machine translation systems to get confused or make mistakes (Lockwood 2000).

With this in mind, we can point to at least two instances where the translation circuit in *Doctor Who* behaves contrary to what we would expect from a machine translation system. In one case, the translation circuit was unable to handle Esperanto, an artificially constructed language known for having a simple and highly regular grammar (Audio story: "A Thousand Tiny Wings," 2010). Meanwhile, in another instance, the language Judoon was considered too basic to be translated by the translation circuit ("Smith and Jones," 2007; Audio story: "Judoon in Chains," 2016). In principle, these comparatively simple languages would actually be easy for a machine translation system to handle, meaning this is an area where *Doctor Who* contradicts the reality of how machine translation actually works.

Finally, an appealing translation circuit characteristic is the "swear filter," which bleeps out offensive language (*Only Human*, 2005). Sadly, this is another area of dissimilarity between the translation circuit and machine translation. As described earlier, corpus-based approaches to machine translation require training data in the form of previously translated texts. A neural machine translation system uses these examples to "learn," but sometimes it learns things that we would rather it didn't. For instance, a recent concern is that neural machine translation systems could suffer from *machine bias* or *algorithmic bias*, meaning that they can perpetuate different types of bias (e.g., racial, gender, age) if the training data are not well selected (Prates et al. 2019). For instance, current neural machine translation systems have a tendency to perpetuate—or even exaggerate—a male bias (Vanmassenhove et al. 2018: 3003). As these researchers explain, human translators rely on contextual information to infer the gender of the speaker in order to make the correct morphological agreement (e.g., translating "I am happy" into French as either "*Je suis heureux*" or "*Je suis heureuse*" according to whether the speaker is male or female). However, most current machine translation systems do not take context into account but simply exploit statistical dependencies on the sentence level that have been learned from large volumes of parallel corpora. Moreover, each sentence is translated in isolation. As a result, information necessary to determine a speaker's gender might get lost, and the machine translation system will select the option that is statistically the most likely variant. Hence the importance of using training corpora that are not skewed or biased. This issue was picked up by the popular media, who reported evidence that Google Translate produced translations that would generally skew toward masculine pronouns for words like "strong" or "doctor" and feminine ones for "beautiful" and "nurse." Google later published two blog posts outlining their efforts to address the problem (Johnson 2018; Kuczmarski 2018). Similarly, there are reports of mistranslations that promote certain political agendas over others, for instance, Google Translate reportedly mistranslated the phrase "I am *sad* to see Hong Kong become part of China" to "I am *happy* to see Hong Kong become part of China" amid pro-democracy protests in Hong Kong in June 2019 (Klar 2019). An apparent example of language in *Doctor Who* being censored in such a way that it promotes a political agenda or norm occurred when Clara was trapped inside a Dalek in "The Witch's Familiar" (2015). While, presumably, some form of Dalek translation technology was translating her, not the TARDIS translation circuit, her words were rendered only in a way that would be "politically" acceptable to those from Skaro. As trapped Clara uttered words like "I love you" and "You are different from me," she was horrified to hear only "Exterminate!" be rendered by the Dalek.

Concluding Remarks

While we appreciate *Doctor Who*'s entertainment value, when it comes to its presentation of translation technology, the series does something of a disservice to both the general public and the translation industry. On a basic level, presenting the translation circuit as a *translation*, rather than an *interpretation*, device reinforces the public's long-standing yet erroneous tendency to conflate these two distinct disciplines. Meanwhile, the suggestion that the translation circuit has difficulty with simplified languages flies in the face of everything we know about machine translation, which finds simplified languages much easier to process.

As pointed out by Bowker (2020), the popular media more or less equate machine translation with science fiction, often invoking a reference to *Star Trek*, the *Hitchhiker's Guide to the Galaxy*, or *Doctor Who*, when discussing machine translation. What's more, machine translation systems are frequently the target of jokes in the media, where typical headlines pulled from a database of major Canadian daily newspapers include "Computers can't tell avocados from lawyers" or "Translation software can yield laughable results." In many cases, machine translation systems are mocked for being unable to translate Christmas carols, fortune cookies, or poetry, even though they were never designed nor intended to process such texts. In similar fashion, *Doctor Who*'s translation circuit is used to generate some humorous moments, such as the conversation between the Doctor's companion Clara and a Russian-speaking character named Grisenko ("Cold War," 2013):

> GRISENKO: I am hungry like the wolf.
> CLARA: I'm not singing.
> GRISENKO: Don't you know it?
> CLARA: Course I know it. We do it at karaoke, the odd hen night.
> GRISENKO: Karaoke? Hen night? You speak excellent Russian, my dear, but sometimes I don't understand a word you're talking about.

Humor arises here because the translation circuit, like a machine translation system, sometimes translates phrases compositionally without accurately conveying the underlying meaning of idiomatic expressions, such as "hen night," which has nothing to do with hens but is a reference to a party for a bride-to-be that might involve going out to a nightclub and singing. A literal translation would therefore fall flat—or generate a laugh, depending on your point of view!

Nevertheless, apart from the odd translation bumble used for humorous effect, science fiction series tend to make automatic translation appear amazingly easy—almost magical—thus diminishing the hard work that has taken place behind the scenes in real life. Machine translation in such series is perhaps best described not as "science fiction," but "science fantasy." Kenny (2011) identifies this type of portrayal of machine translation as something of an ethical problem, noting that, in addition to relying on human translations for training data, corpus-based machine translation systems also rely on human translation for their legitimacy. The reason developers train machine translation systems with parallel corpora is because these corpora of human translations are assumed to contain good solutions. So what is the ethical issue here? The problem is that the contributions made by humans to the process of machine translation go largely unnoticed. As described by Kenny (2011: 2) this technology "relies on the ingenuity of both human translators (who produce vital data) and statistically-minded computer scientists (who work out clever ways of using these and other data), and both sets of protagonists might expect to be acknowledged in discussions." In other words, there is nothing magical about machine translation, which has evolved through years of hard work rather than simply bestowing it as a convenient gift.

Finally, while the idea of the swear filter is charming, it masks what is potentially one of the most insidious features of recent approaches to machine translation: the notion that these systems could suffer from machine bias—or even be explicitly manipulated—if their training material is not carefully selected.

In short, when it comes to the science of translation technology, *Doctor Who* gets it wrong more often than it gets it right. However, perhaps we can forgive the artistic license if we recognize that, as in other science fiction works, the presentation of some type of ubiquitous translation tool is necessary to explain to the audience how people from other countries, time periods, and even other worlds, can understand each other and indeed appear to speak (mostly) flawless English. If this premise were not firmly in place, most other elements of the series simply would not work.

Notes

1. Though apparently very badly, because of Tom Baker's lack of training in the language (Orman, 2013).
2. Although we may rightly critique *Doctor Who* for drawing parallels between communication *systems* like semaphore to British Sign Language and other natural human languages, ironically, the translation circuit does mirror today's machine translation to some extent. Despite numerous attempts at machine translation that incorporate signed languages, no existing system has much practical use. For example, in a viral video demonstrating machine translation for American Sign Language to English, the inventor of the SignAloud system is wearing hand-tracking gloves and signing according to English rather than American Sign Language grammar (University of Washington 2016). It seems science has a long way to go before machine translation for signed languages becomes anything more than a novelty.

References

ALPAC (1966) *Language and Computers in Translation and Linguistics. A Report by the Automatic Language Processing Advisory Committee*. Washington, D.C.: National Academy of Sciences, National Research Council Publication 1416. www.mt-archive.info/ALPAC-1966.pdf.

Baker, M. (2005) Narratives in and of translation. *SKASE Journal of Translation and Interpretation* 1(1): 4–13.

Bowker, L. (2020, April 16) From science fiction to reality: Representations of machine translation in the Canadian press. *Circuit* 146.

Bulis, C. (1996) *Doctor Who: The Eye of the Giant*. Virgin Publishing.

Bulis, C. (2000) *Doctor Who: Imperial Moon*. BBC Worldwide Ltd.

Dalvi, F., Zhang, Y., Khurana, S., Durrani, N., Sajjad, H., Abdelali, A., Mubarak, H., Ali, A., and Vogel, S. (2018) QCRI's live speech translation system. *Qatar Foundation Annual Research Conference Proceedings, 18*(3). Doha: Hamad Bin Khalifa University Press. doi:10.5339/qfarc.2018.ICTPD405.

Esselink, B. (2000) *A Practical Guide to Localization*. Amsterdam/Philadelphia: John Benjamins.

Federici, F., and Al Sharou, K. (2018) Moses, time, and crisis translation. *Translation and Interpreting Studies* 13(3): 486–508.

Hutchins, W.J. (ed.) (2000) *Early Years in Machine Translation: Memoirs and Biographies of Pioneers*. Amsterdam/Philadelphia: John Benjamins Publishing.

Hutchins, W.J. (2004) Two precursors of machine translation: Artsrouni and Trojanskij. *International Journal of Translation* 16(1): 11–31.

Johnson, M. (2018, December 10) Providing gender-specific translations in Google Translate. Blog post on Google AI Blog. ai.googleblog.com/2018/12/providing-gender-specific-translations.html (accessed 3 September 2019).

Kenny, D. (2011) The ethics of machine translation. In: Ferner, S. (ed.) *Reflections on Language and Technology—The Driving Forces in the Modern World of Translation and Interpreting: Proceedings of the 20th New Zealand Society of Translators and Interpreters National Conference*, 4–5 June 2011. Auckland, New Zealand: NZSTI, 121–131. doras.dcu.ie/17606/.

Kenny, D. (2018) Machine Translation. In: Rawling, P., and Wilson, P. (eds.) *Routledge Handbook of Translation and Philosophy*. London: Routledge, pp. 428–445.

Klar, R. (2019, June 18) Google under fire for mistranslating Chinese amid Hong Kong protests. *The Hill*. https://thehill.com/policy/international/asia-pacific/449164-google-under-fire-for-mistranslating-chinese-amid-hong-kong (accessed 3 September 2019).

Koehn, P. (2010) *Statistical Machine Translation*. Cambridge: Cambridge University Press.

Kuczmarski, J. (2018, December 6) Reducing gender bias in Google Translate. *The Keyword*. blog.google/products/translate/reducing-gender-bias-google-translate/ (accessed 3 September 2019).

Langlais, P., Gandrabur, S., Leplus, T., and Lapalme, G. (2005) The long-term forecast for weather bulletin translation. *Machine Translation* 19(1): 83–112.

Lockwood, R. (2000) Machine translation and controlled authoring at Caterpillar. In: Sprung, R. (ed.) *Translating into Success: Cutting-edge Strategies for Going Multilingual in a Global Age*. Amsterdam/Philadelphia: John Benjamins, pp. 187–202.

Munday, J. (2016) *Introduction to Translation Studies: Theories and Applications*, 4th edition. London: Routledge.

O'Brien, S., and Cadwell, P. (2017) Translation facilitates comprehension of health-related crisis information: Kenya as example. *JoSTrans: The Journal of Specialised Translation* 28: 23–51.

Orman, K. (2013) "One of us is yellow": Doctor Fu Manchu and *The Talons of Weng-Chiang*. In: Orthia, L. (ed.) *Doctor Who and Race*. Bristol: Intellect, pp. 83–99.

Orthia, L. (2013, May 23) A very good googly—race in Four to Doomsday. *Doctor Who and Race*. doctorwhoandrace.com/2013/05/23/a-very-good-googly-race-in-four-to-doomsday/ (accessed 5 December 2019).

Pöchhacker, F. (2004) *Introduction to Interpreting Studies*. London: Routledge.

Poibeau, T. (2017) *Machine Translation*. Cambridge: MIT Press.

Prates, M., Avelar, P., and Lamb, L. (2019) Assessing Gender Bias in Machine Translation: A Case Study with Google Translate. *Neural Computing and Applications*. Prepublished 27 March. doi:10.1007/s00521-019-04144-6.

University of Washington (2016, April 12) UW undergraduate team wins $10,000 Lemelson-MIT Student Prize for gloves that translate sign language. www.washington.edu/news/2016/04/12/uw-undergraduate-team-wins-10000-lemelson-mit-student-prize-for-gloves-that-translate-sign-language/ (accessed 3 September 2019).

Vanmassenhove, E., Hardmeier, C., and Way, A. (2018) Getting gender right in neural machine translation. *Proceedings of the 2018 Conference on Empirical Methods in Natural Language Processing*. Brussels, Belgium, October 31-November 4, 2018, 3003–3008.

Way, A. (2019) Machine Translation: Where are we at today? In: Angelone, E., Ehrensberger-Dow, M., and Massey, G. (eds.) *The Bloomsbury Companion to Language Industry Studies*. London: Bloomsbury, 311–332.

Weaver, W. (1949) *Translation*. www.mt-archive.info/Weaver-1949.pdf.

Wen, S. (2019) Translation analysis of English address image recognition based on image recognition. *Journal on Image and Video Processing* 2019: A11. doi:10.1186/s13640-019-0408-9.

"I don't want to go"
How Does Regeneration Work in Doctor Who?

NATALIE RING

Introduction: Regeneration in Doctor Who

Doctor Who premiered on the BBC in November 1963 with the serial "An Unearthly Child." For three years, the Doctor was a grandfatherly figure, played by William Hartnell. During these three years, *Doctor Who* steadily grew in popularity, particularly after the 1964 introduction of the Doctor's archnemeses, the Daleks. By 1965, the show regularly drew upwards of 9 million viewers in the United Kingdom alone (Marlborough 1964). However, by 1966, Hartnell's ill health necessitated a change in lead actor. On screen in "The Tenth Planet" (1966, Hartnell's final story as lead actor) the Doctor grows increasingly frail and aged, and comments, "this old body of mine is wearing a bit thin." In the next story, "The Power of the Daleks" (1966), the Doctor eloquently describes the process that has taken place, comparing his drastic changes in appearance to a butterfly bursting from the chrysalis, and reminding his companions, "life depends on change and renewal." This in-universe comment also speaks to the production context, in which the survival of *Doctor Who* depended on a new actor taking over the role but keeping the audience. To minimize the jarring impact of this change, *Doctor Who* producer Innes Lloyd and story editor Gerry Davis, came up with an in-universe method of explaining it: a biological process which was originally termed "renewal," but would eventually go on to be known as "regeneration." When faced with extreme old age or a fatal injury, the Doctor's people (later identified as the Time Lords when the Troughton incarnation was placed on trial in the 1969 serial "The War Games," although more recently "The Timeless Children" [2020] has raised questions about this) can heal themselves. This process also results in a change in appearance and, to a greater or lesser extent, personality.

According to the Ninth Doctor himself in "The Parting of the Ways" (2005): "Time Lords have this little trick, it's sort of a way of cheating death. Except … it means I'm gonna change, and I'm not gonna see you again. Not like this, not with this daft old face."

The necessary change in their first lead actor forced the BBC to come up with a process which would ultimately help *Doctor Who* to remain on our screens for (at the time of writing) almost 60 years. They had, as Michael G. McDunnah (2019) recently wrote for *Conscious Style Guide*, "stumbled upon one of the greatest and most useful gimmicks in television history." Regeneration has enabled the show to continue despite numerous changes to the actor playing the Doctor, as well as allowing periodic refreshes to format and supporting cast to maintain the interest of contemporary audiences. The Doctor's regeneration has often coincided with major changes in the writers, producers and showrunners of *Doctor Who*. For example, the change from the Second to the Third Doctor in 1969–70 and the Third to the Fourth in 1974 coincided with changes to the producer from Derek Martinus to Barry Letts and Letts to Philip Hinchcliffe respectively. The regeneration of the Tenth Doctor into the Eleventh in 2010 coincided with a handover of the showrunner position from Russell T Davies to Steven Moffat, while the regeneration of the Twelfth Doctor into the Thirteenth in 2017 coincided with Moffat's exit and the arrival of Chris Chibnall as head writer. Regeneration has, therefore, helped to lessen the impact in any changes of style or tone that have resulted from a change in production staff; any major differences in the newer episodes can be seen simply as a consequence of the Doctor's recent change in personality, thus helping to maintain continuity. As of 2020, the Doctor has been played by (at least) fifteen different actors (including John Hurt's War Doctor in "The Day of the Doctor," 2013, and Jo Martin's mystery Doctor in the 2020 season). Over its 57 (and counting) year run, *Doctor Who* has given many examples of, and explanations for, the regeneration process. In this essay, I will explore and expand on some of these,[1] with a particular focus on the post–2005 revival series (during which, the rules and appearance of regeneration have remained relatively consistent compared to the greater diversity in the change process seen in the classic series), using examples of similar processes we see in nature, to attempt to explain how regeneration works in *Doctor Who*.

Time Lord Regeneration: A Few Rules

As may be expected from a show that has had more than one hundred different writers, working under different production regimes, the information given about regeneration has not always been consistent, including the name of the process itself. The appearance of the process also lacked consistency, until the post–2005 series. It is clear, though, that the process though is explosive, traumatic and disorienting for the Time Lord at the center of it, and in the process vast amounts of energy are expended. For this essay, six basic rules for Time Lord regeneration have been defined:

1. Despite their change in appearance and personality, Time Lords retain their memories after regeneration

The Doctor, for example, has always retained their memories of events that occurred during previous regenerations and, as evidenced by multi-Doctor stories such as "The Three Doctors" (1972–73), "The Five Doctors" (1983), "The Two Doctors" (1985), "The Day of the Doctor" and "Twice Upon a Time" (2017), is able to recognize and interact with previous incarnations of themselves.

An exception to this rule may be the Thirteenth Doctor. In "Fugitive of the Judoon" (2020), Jodie Whittaker's Doctor meets a mysterious character, played by Jo Martin, who claims to an older version of the Doctor we have not met before. Neither the Thirteenth Doctor nor Martin's Doctor appears to have any recollection of the other. However, this lack of memory is seemingly explained in "The Timeless Children," when the Master tells the Doctor she has had her memories tampered with.

2. Regeneration is a random process

According to the Ninth Doctor, the regeneration process is "a bit dodgy," and "[y]ou never know what you're gonna end up with." In the vast majority of regenerations we have seen on screen, of the Doctor, Master, and certain other Time Lords, the process has indeed been random and uncontrolled. Nonetheless, a few exceptions to this rule do exist. In "The Stolen Earth/Journey's End" (2008), the Tenth Doctor (later described by the Eleventh as having had "vanity issues") chooses to use regeneration energy to heal from being shot by a Dalek, but "stop[s] the energy going all the way." He thereby prevented himself from changing appearance, which suggests that a Time Lord is able to choose not to regenerate fully (assuming a receptacle for the spare regeneration energy is nearby).

In "The Night of the Doctor" (2013), a one-off mini episode made available online in the build-up to the release of the 50th anniversary special "The Day of the Doctor," the Sisterhood of Karn offer the Eighth Doctor a choice of who he regenerates into, through using an elixir. Ultimately, this elixir does appear to work, as the Eighth Doctor asks to become a warrior, and subsequently regenerates into the War Doctor. Likewise, when the Time Lords force the Second Doctor to regenerate into the Third at the end of "The War Games" (1969), they offer him a variety of different appearances to choose from, although it is unclear how they would control the process.

Finally, in "Destiny of the Daleks" (1979), Romana appears to have the ability to change her appearance at will several times during regeneration, ultimately choosing to look the same as a woman she had previously met (Princess Astra of Atrios, in "The Armageddon Factor," 1979). This ability is never explained, however, and we have never again seen a Time Lord so specifically select their appearance.

Overall, in the post–2005 series, if not the classic, it has been made quite clear that each Time Lord has little control over what their future self will be like, especially without external help.

3. Regeneration is followed by a period of amnesia, recuperation and exhilaration

Often referred to as "rebooting" (for example, by the Twelfth Doctor in "Deep Breath" [2014] and the Thirteenth Doctor in "The Woman Who Fell to Earth," 2018), the first few hours after a regeneration can be turbulent. The Doctor is usually missing some memories during this period; the Thirteenth Doctor cannot remember her own name in "The Woman Who Fell to Earth," while in "Deep Breath" the Twelfth Doctor cannot remember his companions:

> THE DOCTOR: (To Clara) I remember you. You're Handles. You used to be a little … a little robot head, and now you … you've really let yourself go.
> THE DOCTOR: (To Strax) Come on Clara, you know that I speak dinosaur…. Well, you're very similar heights! Maybe you should wear labels?

The Doctor is also prone to collapsing during the first few hours after a regeneration. The Tenth Doctor collapses several times in "The Christmas Invasion" (2005), saying he is having "a neuron implosion." In "Deep Breath," the Twelfth Doctor also passes out repeatedly, although he resists rest. This resistance of rest is characteristic of another trait the Doctor often possesses in the aftermath of a regeneration: over-excitement, exhilaration, and more than a slight hint of silliness. The Twelfth Doctor flirts with a dinosaur, the Eleventh Doctor experiments with a variety of foods only to discover that his favorite is fish fingers dipped in custard, and the Tenth Doctor accidentally quotes a Disney classic during what is supposed to be a serious speech:

> Look at these people. These human beings. Consider their potential. From the day they arrive on the planet and blinking step into the sun, there is more to see than can ever be seen. More to do than … no, hold on. Sorry, that's The Lion King. But the point still stands. Leave them alone!

4. Regeneration is associated with large amounts of energy, and is related to exposure to the Time Vortex

In "The Parting of the Ways," the Ninth Doctor's regeneration is triggered after he absorbs Time Vortex energy from Rose Tyler; the Time Vortex energy has the same yellowy fire we now associate with regeneration. Regeneration energy can also be destructive; in "The Time of the Doctor" (2013), the Eleventh Doctor is able to use his regeneration energy to destroy an attacking Dalek fleet, while the violent regeneration of the Twelfth Doctor in "Twice Upon a Time" causes extensive damage to the TARDIS.

In "A Good Man Goes to War" (2011), we hear that Time Lords "became what they did" after billions of years of exposure to the Time Vortex. This is seemingly directly contradicted in "The Timeless Children," when we learn that the genes for regeneration were in fact stolen from elsewhere and spliced into the Gallifreyan genome. Nonetheless, in "The Sound of Drums" (2007), the Doctor reveals that part of being initiated into the Time Lord academy involves looking into the Untempered Schism, "a gap in the fabric of reality from which can be seen the whole of the Vortex." In addition, we are told in "A Good Man Goes to War" that River Song, a human, is able to regenerate due to the presence of "Time Lord" DNA in her own DNA, thought to have resulted from her being conceived during flight through the Time Vortex. This supports both the idea that regeneration could be achieved by adding regenerative genetic material to the existing DNA of another species, and that the Time Vortex is important in the process.

5. There is a limit to the number of times a Time Lord can regenerate

This is confirmed in "The Time of the Doctor" by the Eleventh Doctor, who believes his current incarnation to be his last, thanks to the regeneration of the Eighth Doctor into the War Doctor before the Ninth Doctor in "The Night of the Doctor," and the aborted regeneration of the Tenth Doctor in "The Stolen Earth/Journey's End" (2008). In "The Timeless Children," we learn that the creator of the Time Lords deliberately imposed this limit.

The Time Lords, however, seem to be able to grant new regeneration cycles. In "The Five Doctors," the High Council of Time Lords appear to offer this to the Master, suggesting they will give him a "complete new life cycle" in return for rescuing the Doctor. Later, in "The Time of the Doctor" we see what seems to be regeneration energy sent to the Doctor from the Time Lords through a time crack, ultimately allowing him to regenerate for a fourteenth time, into the Twelfth Doctor. Events in "The Timeless Children" raise the question of whether the Doctor truly does have a limited number of regenerations, but at the same time confirm again that other Time Lords definitely do.

6. **A previously male Time Lord can regenerate into a female, and vice-versa**

The most recent example is the Twelfth Doctor's regeneration into the Thirteenth in "Twice Upon a Time." However, we had already seen other examples of Time Lords appearing to become a different sex following a regeneration, such as Missy, a female incarnation of the Master. In "World Enough and Time" (2017), the Twelfth Doctor tells Bill that Time Lords are "billions of years beyond [our] petty human obsession with gender and its associated stereotypes."

There are, however, some suggestions that individual Time Lords can be prone to being either male or female. For example, after the General regenerates in "Hell Bent" (2015), she says that she had never been a man prior to her last body. Likewise, in "The Doctor's Wife" (2011), the Eleventh Doctor refers to the Corsair in a way that suggests that they were male more often than female: "Fantastic bloke. He had that snake as a tattoo in every regeneration. Didn't feel like himself unless he had the tattoo. Or herself, a couple of times."

These points suggest that in-universe, there is at least some consistency in cause, process and outcomes when a Time Lord regenerates. The following sections of the essay now turn from the fiction (and the production decisions influencing what appears on screen) to different realms of scientific fact, where actual types of regeneration take place.

Regeneration in Nature

Regeneration in Plants

Examples of organisms regenerating are abundant in nature. The most obvious of these is plants, many of which have an extremely high capacity for regeneration. Just like Time Lords, some plants can live and grow for extremely long periods of time. Most of the oldest organisms on our planet are plants, like the oldest known individual tree, a bristlecone pine in the White Mountains of California named "Methuselah" which is believed to be around 5,000 years old. Still older are the 100-acre "Pando" forest in Utah and a ten-mile long underwater sea grass forest in the Mediterranean Sea near Spain. Both of these consist of what appear to be individual plants (50,000 of them, in the case of Pando) which are actually genetically identical clones which share a root system and grew from the same original cells (Stenger 2019). Pando, a quaking aspen tree, is believed to be around 80,000 years old, while the Mediterranean forest of Neptune's grass is believed to be a huge 200,000 years old. To survive for such long periods, the plants must continually regenerate, to replace cells that have become old or damaged. Unlike Time Lords who, we have seen, can eventually succumb to old age, these plants may have the ability to live indefinitely, as long as they are not cut down or afflicted by any diseases along the way.

In the early twentieth century, Gottlieb Haberlandt, an Austrian botanist, set out to discover just how far a plant's capacity to regenerate would stretch (Haberlandt 1902). He believed that by taking mature plant cells and treating them with the appropriate hormones and nutrients, whole new plants could grow. Ultimately, Haberlandt's experiments were unsuccessful, likely because contemporary understanding of plant growth hormones was limited. However, Haberlandt's work paved the way for the development of one of the cornerstones of modern biology: cell culture. In cell culture, small numbers of cells can be nurtured and cultivated into larger numbers of cells, which are then useable in experiments. The methods by which plant cells can be cultured indefinitely were developed over the course of the twentieth century, with important breakthroughs in the 1950s, 60s and 70s. In 1957, for example, the precise combination of plant hormones (auxin and cytokinin) required to stimulate regeneration was discovered (Skoog and Miller 1957). Later work would go on to show that even a single mature plant cell can be coaxed back into an embryonic state, to produce a whole new population of identical plant cells (Backs-Hüsemann and Reinert 1970; Steward et al. 1958).

Under less drastic circumstances, such as when an amateur gardener wants to propagate their favorite Christmas Cactus, whole new plants can be grown from small cuttings of a mature plant, as long as the cutting includes part of a stem; in this way, a single mature plant could give rise to any number of genetically identical new plants via regeneration. This process is reminiscent of the moment in "Journey's End," when a duplicate Tenth Doctor grows from the original's hand, which was cut off in "The Christmas Invasion." In plants, the ability stems(!) from the fact that adult plants contain a type of tissue known as meristem. Sections of meristem are dispersed throughout the plant in any place where growth can happen, for example at the tips of the roots, or the base of leaves. Within the meristem are pools of meristematic cells, from which any type of cell required in that part of the plant can grow (van den Berg et al. 1997). Such cells, which have not yet become specialized for any mature function, are known as undifferentiated. When a plant is damaged, cut, or simply needs to grow, the meristem is used to generate new cells, which become differentiated into whichever cell type (e.g., leaf, root) is required. Once a cell is differentiated, it can usually only give rise to more cells of the same type. Sometimes, however, the meristem itself can be damaged, and the pool of meristematic cells is lost. Under such circumstances, plants possess a regeneration failsafe, a way of regenerating the meristem. Sometimes, plants achieve this by maintaining a kind of secondary meristem, from which the original meristem can be regenerated directly. Alternatively, plants are capable of "dedifferentiating" mature plant cells and reprogramming them to become other types of cells, with other functions (Ikeuchi et al. 2016). Time Lord regeneration could be reminiscent of any of these methods of regeneration; are we therefore to assume that Time Lords are actually alien plants, with their own meristems? Or do we see examples of regeneration elsewhere in nature?

Regeneration in Other Organisms

Unlikely as it may seem, there do exist other types of organism that are as good at regenerating as plants are. The first of these was discovered over 275 years ago, by Abraham Trembley (Trembley et al. 1744). When he discovered a 10 millimeter-long organism in pond water, Trembley wanted to test whether it was a plant or some kind of animal. Assuming that only plants were able to regenerate into new organisms when cut into smaller pieces, he sliced the organism in two. The two sections began to regrow into new organisms that same day, seemingly indicating that they were indeed plants. However, on closer inspection, Trembley saw that both of the newly grown organisms had heads and legs! Trembley had discovered *Hydra*, a type of minuscule swimming polyp, now considered to be a "superstar" of animal regeneration (Bosch 2007). Like plants, *Hydra* seem to be able to regenerate entirely from just a small number of cells. Also like plants (and Time Lords), they age slowly and may even be capable of living and regenerating indefinitely (Boehm et al. 2012; Martinez 1998). The *Hydra* regeneration mechanism is reminiscent of that of a plant: *Hydra* possess a type of epithelial tissue which, like a meristem, consists of cells which are able to differentiate into a wide range of specialized mature cells, according to which type of cell is required. Although not stem cells in the classic sense,[2] *Hydra*'s epithelial cells are described as having "stem cell properties," and a *Hydra* cannot regenerate from any other type of tissue (Bosch 2007). Another form of tiny aquatic creature, "immortal jellyfish" *Turritopsis dohrnii*, demonstrates a method of regeneration similar to that used by plants when their meristem is damaged or destroyed. Like a plant that can dedifferentiate mature cells back into their non-specialized state, adult *T. dohrnii* can undergo "reverse development": they can revert back into their immature larval state, and essentially begin their life-cycle again. This ability means they are able to heal from some damage, and they do not seem to age, meaning they could potentially live forever (Martell et al. 2016).

Evidence exists, then, that Time Lords may not necessarily be plants after all. However, it makes sense that tiny creatures like the *Hydra* and the immortal jellyfish, both smaller than 1 centimeter long, might be able to regenerate fully, as they have a relatively small number of cells to regenerate. Regeneration of a much larger organism, like a Time Lord, would surely be much more complicated, as it would involve the regeneration of so many more cells. The incredible regenerative powers of the Mexican axolotl (*Ambystoma mexicanum*), a type of salamander which can repeatedly regrow limbs, tail, spinal cord, and more, are therefore potentially more relevant here. Although an axolotl is still much smaller than a human (or humanoid Time Lord), regrowing a whole limb requires the regeneration of many orders of magnitude more cells at the same time than the regeneration of a *Hydra* or *T. dohrnii*. Axolotl regeneration has been extensively studied, because of its potential applications to medicine, such as being able to repair a damaged human spinal cord. Consequently, we know that the mechanism an axolotl uses to regenerate is most similar to the regeneration mechanism of the immortal jellyfish, or a plant that has lost its meristem. Mature, specialized cells that surround an injury are coaxed back into their juvenile undifferentiated state, from which they can become whichever type of cell is required to fix the injury (Manly 2011).

To understand how this works, we need a better understanding of what occurs at a molecular level. Although we have been able to figure out a few of the genes which are involved in the process, this research has been hampered by the size of the axolotl genome; at 32 billion DNA letters, the genome of an axolotl is over ten times larger than a human genome, making it difficult and expensive to sequence. New DNA sequencing technologies recently developed have already helped greatly in this area, and may yet enable us to understand exactly how an axolotl regenerates so effectively and, potentially, how we could recreate this regeneration in a human (Nowoshilow et al. 2018). Nonetheless, from what we know already, we may guess that Time Lord regeneration would require the deprogramming and re-specialization of the Doctor's existing cells, rather than growth of new ones.

Even humans can regenerate to some extent. Throughout our lives, our cells age and are replaced by new ones. We see this in our ability to regrow damaged skin, or replace blood cells that have been lost. Certain parts of our bodies are better able to regenerate than others, like our liver, which contains its own supply of stem cells. However, in all of the cases discussed so far, including humans, the regenerated organism has always been a clone of the original. That is, the DNA in the regenerated cells is the same as the DNA in the pre-regeneration cells, and the regenerated cells will therefore appear identical to the pre-regeneration cells. In this way, none of the examples we have seen are similar to Time Lord regeneration that includes as one of its main cornerstones a change in appearance, personality or even sex. Such changes surely indicate that Time Lord regeneration involves large changes in DNA. So, can an organism change its DNA, and how might this work during Time Lord regeneration?

Regeneration and DNA

Given the common understanding of DNA as the instruction manual for life (our DNA is what makes each of us who we are, after all), it may seem counterintuitive that our DNA actually does change constantly throughout our lives. A complete copy of our entire genome can be found in almost every type of cell in our body. In order to regenerate, a cell must therefore make a new copy of all the DNA it contains. The mechanism by which our DNA replicates is not perfect, and small mistakes can happen. Although our cells have ways of noticing these mistakes, some of them may still make it into the new cell. We call these kinds of mistakes in our DNA "mutations," and they take a few different forms. The simplest form is a "single nucleotide polymorphism," or SNP (pronounced "snip"), which is when a single DNA letter is substituted for a different DNA letter. Other types of mutations can be much more complicated, including the insertion, deletion or copying of large or small sections of DNA. Some mutations can cause diseases: for example, most cancers are caused by an accumulation of many mutations in many different genes over time. Other mutations may cause no change at all in the way that a gene behaves, or the effect may only be noticed over several generations. Between two non-related human individuals, it is estimated that there will be around 20 million DNA differences (Auton et al. 2015). While this sounds like a huge number, the typical human genome is actually over 3 billion letters long; every difference between every human on Earth is therefore due to changes in only around 0.5 percent of our DNA. Bearing this in mind, although we (obviously) have no idea how long a Time Lord genome is, it starts to look a little more easy for a Time Lord to completely change their appearance when they regenerate—at least 99.5 percent of their DNA could stay the same!

Mutation is not the only way our DNA changes throughout our life. Epigenetics is the study of ways in which our DNA can change, without any changes to the underlying DNA letter. A clear example of epigenetics that we have already seen is the differentiation of stem cells into a certain cell type. To achieve this specialization, some bits of DNA in the cell are switched off permanently, while others are permanently switched on. A common way for this to happen is by certain chemical markers, like a methyl (CH_3) group, becoming attached to the DNA in specific places.

These markers send signals to the cell, telling it which genes should be active or inactive. Another example of epigenetics is X-chromosome inactivation. Most human females possess two X chromosomes, while most human males possess an X and a Y chromosome. Only one set of X genes is required by either females or males, so females actually possess twice as many copies of each of the genes as they need. This double dose of the X genes could lead to problems, so a process known as "X-chromosome inactivation" happens during embryonic development. During X-chromosome inactivation, one of the X chromosomes is inactivated at random in every cell. The underlying DNA stays the same, ready to be passed on when its owner reproduces, but it is marked with an epigenetic signal (Fang et al. 2019).[3] To some extent, epigenetics in humans is also linked to personality. Identical twins are a fascinating case study for epigeneticists: at conception, each twin possessed exactly the same DNA, but as they grow and develop into adulthood and beyond, many identical twins become less identical. Personality differences between identical twins can be apparent even earlier, as children or younger. A 2008 study of identical twins with different behaviors identified a specific type of epigenetic marker in a specific gene, which was linked with stress response and risk-taking behavior (Kaminsky et al. 2008). A Time Lord's change of personality after regeneration could therefore be related to epigenetic changes in a similar way.

Elsewhere in this collection, Mike Stack considers in depth the recent change from male Twelfth to female Thirteenth Doctor, from both a sociological and biological perspective. Here, however, from a purely genetic perspective, considering how extensively a Time Lord's DNA must change during regeneration, it seems apparent that a Time Lord could change from male to female or vice-versa. As already discussed, humans have two varieties of sex chromosomes: X and Y. The most commonly seen combinations are XX and XY, although other combinations exist (such as XXY, etc.). We have seen that only one active X chromosome is required for the development of a female; if Time Lords were to possess the same set of sex chromosomes as humans, a change from male to female could be as simple as the epigenetic inactivation of the Doctor's Y chromosome. Of course, as fictional characters, we have no idea what a Time Lord's chromosomes would look like. Even human biology is not as simple as XX=female and XY=male, and we should not assume that Time Lord biology would be either. Bird sex chromosomes, for example, function in a different way. The bird sex chromosomes are Z and W; a pair of Zs generally results in a male bird, and ZW in a female (Stevens 1997). Many other organisms on Earth, like sea sponges and plants, are both female and male at the same time, or can readily switch between the two. Given the variety of ways in which we see genetics contribute to sex in Earthly organisms, it therefore seems not only possible than an alien species could appear to change sex, but surprising that we have not seen it happen more often.

A Few More Speculative Theories About Time Lord Regeneration

Even using our knowledge of Earthly regeneration, it is hard to explain how all of the DNA in an organism could change in the same way at the same time, as must happen during Time Lord regeneration. Mutation is an entirely random process; it is highly unlikely that exactly the same mutation will occur in two different cells simultaneously, although every cell that descends from the mutated cell will possess the same mutation. A 2013 study estimated than an average human body contains around 3.72 trillion cells (Bianconi et al. 2013). Assuming Time Lord cells are around the same size as human cells, we may conclude that the body of a Time Lord of average human height (like the Doctor tends to be) would contain a similar number of cells. To change the DNA in every one of these cells at the same time, something would be needed to control the change. At this point, we must veer into speculation entirely, albeit speculation grounded in the above accounts of both plant and animal regeneration. Perhaps each Time Lord possesses a store of stem cells, from which they can entirely regrow (like a plant's meristem). They could even possess several different stores of stem cells (say, thirteen?), each of which contains cells with slightly different DNA. During regeneration, one of the stores would be used, resulting in a change of DNA in every regenerated cell. Alternatively, perhaps Time Lord regeneration is more like the regeneration of an axolotl, during which mature cells are dedifferentiated. While dedifferentiated, perhaps the DNA is changed. This second method would likely require an entirely alien organ, to control the change. It has been well established in *Doctor Who* that Time Lords have two hearts, so it would be possible that they would have other additional physiological differences. Either method explains why regeneration is a random process that can go wrong: just like the replication of human DNA during cell regeneration, mutations could arise and go unfixed. However, neither method explains why the Sisterhood of Karn could offer the Eighth Doctor a choice of who he regenerated into in "The Night of the Doctor."

Whether every cell in a Time Lord's body has its DNA changed simultaneously, or whether every cell spontaneously dedifferentiates and regrows with different DNA, regeneration is a process that would require a huge amount of energy. We are still in the land of pure speculation, but it is possible that Time Lords' exposure to the Time Vortex, which presumably abounds with timey-wimey energy (often referred to as "Artron" energy, for example in "Spyfall," 2020), provides them with the energy they need to regenerate. However, this very science fiction (rather than science fact) theory would imply the existence of another special Time Lord organ, one that could store regeneration energy for when it is needed.

One final aspect of Time Lord regeneration which does not seem to make sense according to Earthly biology is the Time Lords' ability to retain their memories when they regenerate. Our understanding of memory is still incomplete, but we do know that many different parts of the brain are involved in the formation of short- and long-term memories, including the hippocampus, cerebellum and amygdala (Frankland and Bontempi 2005). From the moment we are born, we start to form memories. During early development, most of these memories are related to learning: how to eat, walk and talk, for example. The complete regeneration of every cell in a human brain could therefore not only cause that human to forget everything that has happened to them, it could also cause them to revert to an infant-like state, in which they are barely able to function without external help! Clearly, either the cells in a Time Lord's brain do not regenerate with the rest of their body, or Time Lord memories are stored differently to human memories. The first of these options seems less feasible, because it would result in the Time Lord's brain cells containing different DNA compared to the rest of their body. Our experience with organ transplants indicates that any part of the body containing very different cells compared to the rest could be problematic.

Let us assume, then, that Time Lord memories are stored differently than human memories. Perhaps there exists a section of the Time Lord brain, which is entirely dedicated to the storage of every type of memory. Indeed, the Doctor herself hinted at the existence of a second brain with a throwaway comment in "Praxeus" (2020). Like certain human cells (e.g., red blood cells), perhaps these memory cells do not contain DNA, and would thus not need to change during regeneration. During regeneration, this section of the brain could disconnect, to be reconnected later. Surprisingly, there is some evidence from the post-regeneration episodes of the show that does seem to support this wildly hypothetical theory, such as the Doctor's brief period of amnesia following some regenerations, and the Twelfth and Thirteenth Doctors' references to waiting for their brains to "reboot." Alternatively, perhaps memories are temporarily stored in the Time Lords' Matrix; in "The Deadly Assassin" (1976) we learn that the minds of Time Lords are transferred to the Matrix at their moment of death, and in "The Timeless Children," the Master further describes the Matrix as "A data bank of every Time Lord consciousness, living and dead. Every experience and every memory." The delay between this temporary storage and later restoration of the Doctor's memories could also explain the "reboot" comments.

In any case, a Time Lord's brain goes through a significant trauma during regeneration. This trauma hints at an explanation for the Doctor's often slightly odd behavior in the first day or so after a regeneration. Humans who suffer from seizures sometimes experience "postictal delirium," while those who suffer from migraines may experience a "postdrome." In both cases, the brain has gone through a traumatic event, and a potential symptom of both is euphoria and/or mania (Barbanti et al. 2013; Kaplan 2003). The postdrome euphoria experienced after a migraine is thought to be caused by a surge of the neurotransmitter dopamine throughout the migraine, while the more severe postictal delirium is thought to be related to the widespread disruption of electrical signaling in the brain during a seizure. It results in the dysregulation of many neurotransmitters, including GABA, glutamate and endogenous opiates (Barbanti et al. 2013; Sachdev 2007). Given the disruption that would occur in a Time Lord's brain during regeneration, the Doctor's behavior after regenerating suddenly does not seem so odd!

Conclusion

This essay has explored possible biological features of Time Lord regeneration, occasionally veering from being solidly grounded in science fact towards wildly speculative science fiction. Although these points are speculative, it is possible to explain many of the rules of regeneration the TV show has introduced using our current biological understanding of both plants and animals. Good science fiction (which includes *Doctor Who*) is richly creative and draws creative impulses from the history of science and its innovative knowledge. The production contingencies of the BBC in 1966 and the need to find a way for *Doctor Who* to outlast the ailing William Hartnell, can interlock with the biological lives of both plants and animals. Whether intentional or not, the invention of regeneration as a method of changing lead actor has provided a rich source of scientific discussion. Events in the most recent episodes have added additional layers to this discussion, and as the show progresses in the years ahead, it is safe to assume that it will continue to alter, and expand upon, our understanding of Time Lord regeneration.

Notes

1. Due to the sheer volume of content available (some of it conflicting, or not widely available), for the sake of simplicity, only examples from the TV show and one-off specials will be considered here.
2. The classic definition of a stem cell says they are undifferentiated cells that can become differentiated into other types of specialized mature cells with different functions, and which can also replicate into more stem cells. A number of different types of stem cells exist, with varying abilities to become other cell types. The most flexible of these are the totipotent stem cells, which can become literally any other type of cell in that organism (like the cells found in a very early human zygote, just hours after conception). The least flexible are the unipotent stem cells, which have the ability to replicate indefinitely, but can only differentiate into one type of mature cell (like the unipotent hepatoblast stem cells found in a human liver).
3. The X-inactivation process also explains why we only see female tortoiseshell cats: the genes coding for fur pattern are found on the cat's X chromosome, and the copies of the gene on each of the inherited chromosomes can code for different colors. In each cell in the developing cat embryo, a different copy of the chromosome may be inactivated, so different parts of the cat can have different colored fur. In a male cat, with only one copy of the X chromosome, the gene coding for fur color will be the same in every cell.

References

Auton, A., Brooks, L.D., Durbin, R.M., Garrison, E.P., Kang, H.M., Korbel, J.O., Marchini, J.L., McCarthy, S., McVean, G.A., and Abecasis, G.R. (2015) A global reference for human genetic variation. *Nature* 526(7571): 68–74. doi:10.1038/nature15393.

Backs-Hüsemann, D., and Reinert, J. (1970) Embryobildung durch isolierte Einzelzellen aus Gewebekulturen vonDaucus carota. *Protoplasma* 70(1): 49–60. doi:10.1007/BF01276841.

Barbanti, P., Fofi, L., Aurilia, C., and Egeo, G. (2013) Dopaminergic symptoms in migraine. *Neurological Sciences* 34(1): 67–70. doi:10.1007/s10072-013-1415-8.

Bianconi, E., Piovesan, A., Facchin, F., Beraudi, A., Casadei, R., Frabetti, F., Vitale, L., Pelleri, M.C., Tassani, S., Piva, F., Perez-Amodio, S., Strippoli, P., and Canaider, S. (2013) An estimation of the number of cells in the human body. *Annals of Human Biology* 40(6): 463–471. doi:10.3109/03014460.2013.807878.

Boehm, A.M., Khalturin, K., Anton-Erxleben, F., Hemmrich, G., Klostermeier, U.C., Lopez-Quintero, J.A., Oberg, H.H., Puchert, M., Rosenstiel, P., Wittlieb, J., and Bosch, T.C. (2012) FoxO is a critical regulator of stem cell maintenance in immortal Hydra. *Proceedings of the National Academy of Sciences* 109(48): 19697–19702. doi:10.1073/pnas.1209714109.

Bosch, T.C. (2007) Why polyps regenerate and we don't: Towards a cellular and molecular framework for Hydra regeneration. *Developmental Biology* 303(2): 421–433. doi:10.1016/j.ydbio.2006.12.012

Fang, H., Disteche, C.M., and Berletch, J.B. (2019), X. Inactivation and Escape: Epigenetic and Structural Features. *Frontiers in Cell and Developmental Biology* 7: 219. doi:10.3389/fcell.2019.00219.

Frankland, P.W., and Bontempi, B. (2005) The organization of recent and remote memories. *Nature Reviews Neuroscience* 6(2): 119–130. doi:10.1038/nrn1607.

Haberlandt, G. (1902) Kulturversuche mit isolierten Pflanzenzellen. *Sitz-Ber. Mat. Nat. Kl. Kais. Akad. Wiss. Wien* 111: 69–92.

Ikeuchi, M., Ogawa, Y., Iwase, A., and Sugimoto, K. (2016) Plant regeneration: Cellular origins and molecular mechanisms. *Development* 143(9): 1442–1451. doi:10.1242/dev.134668.

Kaminsky, Z., Petronis, A., Wang, S.C., Levine, B., Ghaffar, O., Floden, D., and Feinstein, A. (2008) Epigenetics of personality traits: An illustrative study of identical twins discordant for risk-taking behavior. *Twin Research and Human Genetics* 11(1): 1–11. doi:10.1375/twin.11.1.1.

Kaplan, P.W. (2003) Delirium and epilepsy. *Dialogues in Clinical Neuroscience* 5(2): 187–200.

Manly, D. (2011) Regeneration: The axolotl story. *Scientific American*. blogs.scientific american.com/guest-blog/regeneration-the-axolotl-story/ (accessed 31 December 2019).
Marlborough, D. (1964, December 28) Dead, but they won't lie down. *Daily Mail*. cuttingsarchive.org/index.php/Dead,_but_they_won%27t_lie_down.
Martell, L., Piraino, S., Gravili, C., and Boero, F. (2016) Life cycle, morphology and medusa ontogenesis of Turritopsis dohrnii (Cnidaria: Hydrozoa). *Italian Journal of Zoology* 83(3): 390–399. doi:10.1080/11250003.2016.1203034.
Martinez, D.E. (1998) Mortality patterns suggest lack of senescence in hydra. *Experimental Gerontology* 33(3): 217–225. doi:10.1016/s0531-5565(97)00113-7.
McDunnah, M.G. (2019, January 16) Doctor He, She, or They? Changing Gender, and Language, in "Doctor Who." *Conscious Style Guide*. consciousstyleguide.com/doctor-he-she-or-they-changing-gender-and-language-in-doctor-who/ (accessed 8 December 2019).
Nowoshilow, S., Schloissnig, S., Fei, J.F., Dahl, A., Pang, A.W.C, Pippel, M., Winkler, S., Hastie, A.R., Young, G., Roscito, J.G., Falcon, F., Knapp, D., Powell, S., Cruz, A., Cao, H., Habermann, B., Hiller, M., Tanaka, E.M., and Myers, E.W. (2018) The axolotl genome and the evolution of key tissue formation regulators. *Nature* 554(7690): 50–55. doi:10.1038/nature25458.
Sachdev, P.S. (2007) Alternating and Postictal Psychoses: Review and a Unifying Hypothesis. *Schizophrenia Bulletin* 33(4): 1029–1037. doi:10.1093/schbul/sbm012.
Skoog, F., and Miller, C.O. (1957) Chemical regulation of growth and organ formation in plant tissues cultured in vitro. *Symposia of the Society for Experimental Biology* 11: 118–130.
Stenger, R. (2019, December 4) Where to see the oldest living things on Earth. *CNN Travel*. edition.cnn.com/travel/article/oldest-living-things/index.html (accessed 30 December 2019).
Stevens, L. (1997) Sex chromosomes and sex determining mechanisms in birds. *Science Progress* 80 (Pt 3): 197–216.
Steward, F.C., Mapes, M.O., and Mears, K. (1958) Growth and Organized Development of Cultured Cells. II. Organization in Cultures Grown from Freely Suspended Cells. *American Journal of Botany* 45(10): 705–708. doi:10.2307/2439728.
Trembley, A., Pronk, C., Schley Jvd and Lyonet, P. (1744) *Mémoires pour servir à l'histoire d'un genre de polypes d'eau douce, à bras en forme de cornes*. A Leide: Chez Jean & Herman Verbeek.
van den Berg, C., Willemsen, V., Hendriks, G., Weisbeek, P., and Scheres, B. (1997)Short-range control of cell differentiation in the Arabidopsis root meristem. *Nature* 390(6657): 287–289. doi:10.1038/36856.

Did the Doctor Change Sex or Change Gender?
Navigating the Sex and Gender Divide in Doctor Who

MIKE STACK

In 2018, *Doctor Who* premiered its first full series with a female Doctor. Although both "Fugitive of the Judoon" and "The Timeless Children" (2020) suggested there were female incarnations prior to the Thirteenth Doctor, from an audience perspective the character's switch of sex was a bold move for the series after 55 years and 12 male leads. Inevitably, Jodie Whittaker's casting attracted much scrutiny and strong reactions. Yet the series evaded an extended examination of this change by shifting the emphasis away from continuity—references to the Doctor's previous incarnations as male were rare in Whittaker's first series in the role—and instead concentrated on the new characters and plotlines. This essay will explore the Doctor's change of sex in terms of both a sociological understanding of gender and a biological understanding of sex, arguing that critical responses to Whittaker's casting are, in part, informed by presumptions about the latter. Yet, while these scientific narratives are privileged, those working in the biosciences have further challenged them. Science does not produce a uniform truth. To resolve such conflicts, the essay will conclude that a paradigm shift to promote inclusivity across academic disciplines can resolve the tensions that have arisen with the Doctor's change of sex.

Introduction

On 16 July 2017, Jodie Whittaker was announced as the Thirteenth Doctor, part of an overhaul of the series production under incoming producer and head writer Chris Chibnall. Whittaker's introductory "reveal" was also posted on YouTube, and at the time of writing has received 3.4 million views. As a rough indication of audience response, the video received 69K "thumbs up" expressing approval, and 41K "thumbs down" expressing disapproval, indicating a sizeable split in the reception of a female Doctor (YouTube, 2017). The day after her announcement, British newspaper *The Guardian* summed up the social media outrage at the casting of a female Doctor as "beyond parody" (*Guardian* Editorial 2017). Of course, the character's change in sex was not unprecedented. Three years earlier, the character of the Master had also changed sex, now called Missy, played by Michelle Gomez. Audience reception of this prior casting change was similarly mixed. Media theorist Lorna Jowett explored responses to this casting, noting its relative novelty in the canon of the television series: "the extremity of views aired in public through the mainstream press and via social media and the internet was exacerbated by the fact that this was not speculation, it was an accomplished story" (Jowett 2017: 172). Her examples range from "old-school fanboy-type" responses critical to a change of sex, to feminist voices already wishing for the Doctor to be female. For Jowett, both positive and negative reactions were constrained by debates about the series' storylines and failed to recognize wider sexism in the entertainment industry. Similarly, media theorist Dene October analyzed fan responses in online forums, claiming that they "demonstrated that disputes can stem from a sense of entitlement: a loyal audience may become hostile and misogynistic, with complaints ranging from issues of character ownership and canon to anxiety about change" (October 2018: 251). October (2018: 245–46) contrasts these responses with a debate from his media design class, who expressed concerns that this sex change failed to address the complexities of transgender identities. While trans identities are indeed varied and complex, the concept of regeneration does provide a useful analogy for gender transition. Science-fiction author Susan Jane Bigelow, in a personal contribution to the fan collection *Queers Dig Time Lords,* uses the introductory post-regeneration episodes as a way of understanding her own gender transition: "I've watched those scenes over and over. They matter to me" (Bigelow 2013: 215).

It is notable then that *Doctor Who* fandom has recently become embroiled in controversies surrounding trans rights. In 2019, the media reported on the dropping of television-series writer Gareth Roberts from an anthology of short stories by BBC Books because of his offensive anti-trans views, which he had disseminated on social media (Flood 2019). In turn, Roberts defended himself online with a statement outlining his position: "I don't believe in gender identity. It is impossible for a person to change their biological sex. I don't believe anybody is born in the wrong body" (Roberts 2019).

The irony of a claim that it is "impossible" to change sex from a writer whose primary credits are from a series where the lead character has changed sex is hard to ignore. To be clear, I don't agree with Roberts for multiple reasons, first and foremost that his dismissal of trans identities overrides the agency and experiences of trans people (such as Bigelow).[1] Yet, it is his statement's slippage from "gender identity" to "biological sex" that interests me here. The former is undermined ("I don't believe") while the latter is upheld as an absolute ("impossible [...] to change"). What we can see in this statement, then, is an epistemological hierarchy between a sociological understanding of gender and a biological one. To understand how this opposition works, and to describe the tensions between the two positions, we need to understand what we mean by "gender" and "sex."

Gender Theory

Gender theory as a discipline may be seen as developing out of second-wave feminism, with the need to understand and account for the role of men in society alongside examining women's status. In 1987, sociologists Candice West and Hans Zimmerman published their famous paper "Doing Gender," which argued that gender identity was something actively done rather than an inherent property of individuals: *"Gender* [...] is the activity of managing situated conduct in light of normative conceptions of attitudes and activities appropriate for one's sex category. Gender activities emerge from and bolster claims to membership in a sex category" (West and Zimmerman 1987: 127)

What we see in West and Zimmerman's paper is a separation of gender from sex; they understand gender as a sociological achievement, based upon conventionalized depictions, to convey our biological sex to each other. In so doing, gender is an act, something that is *done* rather than a stable concept.

This understanding of gender was greatly expanded by cultural theorist Judith Butler in her 1990 monograph *Gender Trouble*. This famous work critiques feminist and psychoanalytic notions of gender identity as constructions unable to free themselves from the discourses they seek to challenge. For Butler, gender is an all-pervading concept; one that conceals its own artificiality through conveying itself as a natural state:

> acts, gestures, and desire produce the effect of an internal core or substance but produce this *on the surface* of the body, through the play of signifying absences that suggest, but never reveal, the organizing principle of identity as a cause. Such acts, gestures, enactments, generally construed, are *performative* in the sense that the essence or

identity that they otherwise purport to express are *fabrications*, manufactured and sustained through corporeal signs and other discursive means [Butler 1990: 185].

At the heart of the organizing principle here is "an obligatory frame of reproductive heterosexuality" (186) which is so ingrained by social conventions that it assumes itself as the default condition. The usefulness here of Butler's elaborate theorization is not merely its criticism of heteronormativity, but its destabilization of sex and gender altogether. Performativity is a broad concept that describes the way all actions, gestures and routines (such as dress, make-up and behavior) contribute to an elaborate act of gender that defines our sex. Sex does not precede gender; gender precedes and produces sex.

Through performativity, we can begin to pay attention to how the Doctor has carefully suggested her new gender without explicitly referring to it. There are, of course, moments in the 2018 season of *Doctor Who* that acknowledge social divisions of sex. For example, in "Demons of the Punjab," the Doctor observes "This is the best thing ever—never did this when I was a man" while attending a Mehndi ritual for a bride, where men are excluded. In "The Witchfinders," the Doctor observes she would not be subject to suspicions of witchcraft and a potentially fatal trial by dunking if she were male. But these moments are explicit; scenes that are self-aware enough to draw attention to gender divisions. Performativity is internalized, hidden, made invisible by its appearance as natural. More suggestive of this is the Doctor's sudden adoption of earrings—no Doctor has previously worn jewelry in their ears, yet she now does. This sudden embracing of earrings contributes to the perception that jewelry is somehow feminine. While her costume acknowledges gender fluidity—her boots are unisex—the three-quarter-length trousers are subtly reminiscent of a skirt. The Doctor previously rarely exposed their legs—only the Eleventh Doctor has done so in the socially acceptable (and particularly masculine) setting of a football game in "The Lodger" (2010)—but now the Doctor has more flesh on display, if only her shins. Again, it is a subtle change, but both draws upon and contributes to expectations in dress.

Another moment where gender is unconsciously naturalized is the leap from crane to crane at the climax of "The Woman Who Fell to Earth" (2018). Following her running jump, the Doctor announces: "These legs definitely used to be longer." This is a brief reference to the Twelfth Doctor's taller frame, but the utterance does further work—adding to the perceptions that women are shorter than men. As gender theorist Raewyn Connell observes:

> Adult males are on average a little taller than adult females, but the diversity of heights within each group is great, in relation to the average difference. Therefore a very large

number of individual women are taller than many individual men. We tend not to notice this physical fact because of social custom [Connell 2009: 52].

Connell notes that social convention more often than not prompts men to pair with women shorter than themselves (and vice versa), producing the resulting belief that men are taller than women. So the Thirteenth Doctor's statement above does three things: it articulates the fact that she is shorter than the Twelfth Doctor; it emphasizes the peril of her leap; and also reinforces the perception about gender differences in heights. *Doctor Who* has perpetuated the latter belief. Notice how, in previous incarnations, the companion tends to be shorter than the Doctor: Mel and Ace are paired with the generally rather short Seventh Doctor; and the above-average-in-height Amy Pond paired with the even taller Eleventh Doctor. Only Barbara Wright bucks this trend, although she is generally coupled with the taller Ian Chesterton.[2]

This is not to say the Thirteenth Doctor simply reinforces gender conventions. The very fact of a female Doctor is subversive of the series' norms. The Doctor's technical adeptness, exemplified by her building of the sonic screwdriver in the opening episode, challenges stereotypes about women's ability in STEM subjects (discussed by Orthia and de Kauwe, this volume). Nevertheless, Whittaker's portrayal is still embedded within the prevailing social discourses of gender, which are unpinned by scientific notions of biological difference—to which we now turn.

Biological Approaches to Sex

West and Zimmerman define sex as "a determination made through the application of socially agreed upon biological criteria for classifying persons as females or males" (West and Zimmerman 1987: 127). Thus, underlying biology—in particular, sex chromosomes and the reproductive organs—provides the basis for classifying two distinct sexes and to produce a binary system of gender. The idea that men and women are inherently different and unchangeable in terms of psychology, strength, and morphology is known as "essentialism." Feminist theorist Elizabeth Grosz has defined the term:

> Essentialism entails the belief that those characteristics defined as women's essence are shared in common by all women at all times. It implies a limit on the variations and possibilities of change—it is not possible for a subject to act in a manner contrary to her essence. [...] Essentialism thus refers to the existence of fixed characteristics, given attributes, and ahistorical functions that limit the possibilities of change and thus of social reorganization [Grosz 1995: 47–48].

Essentialist ideas are founded on biological convictions, which Grosz (1995: 48) further defines as "biologism": "biology is assumed to constitute an unalterable bedrock of identity, the attribution of biologistic characteristics amounts to a permanent form of social containment for women." While Grosz's definitions are written as part of a feminist critique, the same principles apply to both women and men, fixing gender roles in rigidly constrained expectations of behavior.

We can easily find essentialist ideas about men and women circulating in our culture. The title of the 1992 pop-psychology bestseller by counselor Jonathan Grey exemplifies such beliefs, *Men Are from Mars, Women Are from Venus*—here the difference between the sexes is understood to be so great, that they are entirely different species from separate planets. Scientific understanding contributes to this perception. Elsewhere in this volume, Natalie Ring explores how chromosomal changes will contribute to the Doctor's change of sex; molecular biologists have argued for the far-reaching consequences of genetic material. For example, geneticist Dene Hamer and science writer Peter Copeland have argued:

> Every fetus is created without a sex until a single gene switches on and begins a cascade of chemical reactions that turns half of us into males and half into females. The changes affect not only physical characteristics but also mental ones as well. Men are programmed to seek more partners and sexual novelty; women are "serial monogamists," seeking mates who will remain long enough to raise offspring. Women want emotional attachment and financial security not because that's what they are taught but because it helps the species survive [Hamer and Copeland 1999: 9–10].

Writing on the eve of the new millennium, in the climate of the Human Genome Project and the cloning of Dolly the Sheep, such an assertion may have seemed absolute and unassailable. The scale is impressive: a gene activates, resulting in chemical "cascades" (presumably hormones); people are "programmed" in mind and body. Sexual behavior is inherently different, with women's choices responsible for nothing less than the survival of the species. Essentialism is writ large as human destiny. Genetics reveals the simple binary of gender: that the mind and body are predestined by the straightforward division between those with XY chromosomes (male) and those with XX (female).

Surprisingly, genetics is not mentioned in relation to regeneration in the Thirteenth Doctor's opening episode. Instead, the story uses genetics for one of its core plot points, the DNA bombs implanted by Tzim-Sha: "Micro-implants which code to your DNA. On detonation, they disrupt the foundation of your genetic code." Yet, just two lines of dialogue later (one each from Graham and Ryan), the Doctor describes the regeneration process in terms of computing and makes a parallel with mobile phones instead: "I'm not yet who I am. Brain and body still rebooting, reformatting." This is in sharp contrast to previous episodes where DNA is depicted as potent and infectious, such as "Dalek" (2005), where Rose's DNA confers humanity on a Dalek, or "Daleks in Manhattan/Evolution of the Daleks" (2007) where the Daleks transmit their own DNA to create a human-Dalek hybrid army. Given the apparent significance of genetics on sex, and its use as a MacGuffin in previous episodes (and indeed, the same episode), why, then, does "The Woman Who Fell to Earth" conspicuously evade referencing DNA in relation to regeneration and the Doctor's change of sex?

It would seem, however, that genetics is not as all-powerful as Hamer and Copeland initially suggest. They cite the case of Spanish hurdler María Patiño, who was disqualified from the 1985 World Games when sex testing revealed her to have an XY karyotype. Patiño was found to have androgen insensitivity syndrome (AIS), and so, Hamer and Copeland concede "her external genitalia and sex characteristics were those of a woman" (Hamer and Copeland 1999: 168). However, in describing the syndrome, in spite of this blurring of their own definitions of male and female, the authors still default to a binary description using a peculiar analogy:

> her developmental train should have been directed in the male direction. But because there was no receptor to sense androgen, it was a circular track. No external male features were formed, and the train was routed back to the female track [168].

Sex development is reduced to the status of train journey (though the metaphor is jumbled between a circular and female track). Train tracks are linear and facilitate only pre-defined direction of movement; hence the analogy reveals how a limited gender binary is constructed (a "male track" or "female track").[3] However, reviewing types of intersexuality complicates the binary of gender as presented in terms of genetic sex. Developmental geneticist and historian Anne Fausto-Sterling tabulates the common types of intersexuality in her work *Sexing the Body* (2000), including AIS, as well as Congenital Adrenal Hyperplasia (masculine XX karyotype), Gonadal Dysgenesis (affecting XX and XY karyotypes), Turner Syndrome (XO karyotypes), and Klinefelter Syndrome (XXY karyotypes) (Fausto-Sterling 2000: 52). Collating this data, Fausto-Sterling (2000: 51) challenges the biologism of a gender binary, arguing that such conditions cannot be considered rare as they account for 1.7 percent of all births: "At a rate of 1.7 percent, for example, a city of 300,000 would have 5,100 people with varying degrees of intersexual development."

Indeed, Fausto-Sterling had earlier argued for an alternative gender system that comprises five genders, including men and women and:

> the so-called true hermaphrodites, whom I call herms, who possess one testis and one ovary [...]; the male pseudohermaphrodites (the "merms"), who have testes and

some aspects of female genitalia but no ovaries; and the female pseudohermaphrodites (the "ferms"), who have ovaries and some aspects of the male genitalia but lack testes [Fausto-Sterling 1993: 21].

Even allowing for this disruption of a binary gender system, Fausto-Sterling nevertheless acknowledges, "sex is a vast, infinitely malleable continuum that defies the constraints of even five categories." With such complexity in mind, Time Lord regeneration between two genders suddenly looks simplistic.

Doctor Who as a series has occasionally troubled the binary understanding of gender in its presentation of humans and human-like species. For example, "Sleep No More" (2015) features trans actor Bethany Black as "Grunt" 474, however the character is rather underdeveloped with only a few lines of dialogue and, ultimately, killed. In the main, what is noticeable about the series' challenges to the binary is that it often also reinforces it. For example, in "Kinda" (1982), we receive a tantalizing glimpse of sex on Deva Loka through Karuna's assertion that Aris is one of her fathers. When the Doctor queries how many she has, she answers seven, before innocently asking, "Why? How many fathers does a Not-we have?" As the Doctor's confusion suggests, this moment raises many questions. Is Karuna's statement indicating that Kinda eggs require fertilization from multiple gametes, or indeed are there more mothers involved? Or is it just a passing observation on the social conventions of Kinda parenting? Ironically, however, Deva Loka presents a very rigid binary between male and female, even if subverting patriarchy through matriarchal wisdom. Men are without voice and only the female Kinda speak. As Panna asserts, "No male can open the Box of Jhana without being driven out of his mind." This is clearly, then, a society structured around a gender dichotomy. Such contradictions continue in the post–2005 series. In "The Tsuranga Conundrum" (2018), ideas of pregnancy are subverted through the presentation of a male character, Yoss, being pregnant. While the story playfully inverts gendered depictions of fatherhood, Yoss's explanation once more reveals a rigid binary: "Boys give birth to boys, and girls give birth to girls. That's how it is." Why one or the other? Indeed, why identify as either, especially as the gender of birth parent is now negligible? Thus, even when a key signifier of gender—reproductive biology—is troubled, the series retains a commitment to a binary system of gender.

So why do we still consider sex/gender a binary system, even when challenges to such a binary are so visible? I think the answer lies, in part, because the science itself has produced and facilitated a male/female dichotomy based on reproductive anatomy, ultimately reinforcing it. As Fausto-Sterling points out, intersex children are assigned a sex at birth, surgically modified so their bodies conform to male or female:

> Rather than force us to admit the social nature of our ideas about sexual difference, our ever more sophisticated medical technology has allowed us, by its attempts to render such bodies male or female, to insist that people are either male or female [Fausto-Sterling 2000: 54].

Seemingly arbitrary principles are applied: issues such as the ability to urinate standing up and sexual penetration are considered for assessing suitable penis size, alternatively clitoral size is subject to arbitrary aesthetic judgments that result in clitorectomy, clitoral reduction or recession. Any ambiguity is erased, to ensure that babies conform to a gender binary. Fausto-Sterling raises her concerns about the damage such procedures cause, which may result in multiple surgeries, pain, scarring, loss of sensation, humiliation, and ultimately, loss of agency from the affected individuals.[4] Although Fausto-Sterling was writing in 2000, such practice continues today. For example, legal scholar Bernard M. Dickens wrote about the ethical and legal issues concerning such invasive surgical procedures undertaken on those who cannot decide for themselves, and even concealing such surgery from parents: "Predetermining children's futures by such interventions is also liable to require continuing deception regarding their biological and/or genetic inheritance, contrary to ethical expectations of truth telling and legal requirements of informed consent to treatment" (Dickens 2018: 258). This article was published the same year as the Thirteenth Doctor's debut series, nearly two decades after Fausto-Sterling published her concerns.

Returning to the Thirteenth Doctor's change of sex, such a debate exposes less the mechanics of sex change in regeneration (it may be fictional, but it's never depicted as surgical), but instead the strength of the biologism underpinning the binary system of gender. Such biologism inevitably influenced the debates surrounding—and particularly, the objections to—the Thirteenth Doctor's debut (and Missy's too). But clearly, the supposedly objective and dispassionate pursuit of science is not free of influence from—or culpability in producing—such discursive pressures. When reviewing scientific literature on the biological determination of sex, Judith Butler writes that it is not "that valid and demonstrable claims cannot be made about sex-determination, but rather that cultural assumptions regarding […] the binary of gender itself frame and focus the research in sex determination" (Butler 1990: 148). That is, science itself is not free of the discourses it produces, and these shape the research questions and influence the interpretation of results. Socio-medical scientist Rebecca M. Jordan-Young surveyed several decades of research into "brain organization theory," which she defines thus:

> It rests on a very simple idea: the brain is a sort of accessory reproductive organ. […] brain organization theory is used to explain a very wide range of differences related to

gender and sexuality—in humans these include everything from spatial relations, verbal ability, or math aptitude, to a tendency to display nurturing behaviors, to sexual orientation [Jordan-Young 2010: 21–22].

Brain organization theory has concentrated on the influence of hormones, such as testosterone or estrogen, on development, and Jordan-Young's survey highlights the ambiguous results and assumptions made about the influence of these hormones. However, the supposed indicators of masculine or feminine behavior are socially defined and shaped. In her 2010 book *Delusions of Gender*, cognitive psychologist and historian Cordelia Fine reviews experimental work on subjects exposed to images of gender stereotypes against those who are not, revealing how damaging such negative images are on an individual's self-belief: "these all interact with, and shape, our minds" (Fine 2010: 53). Research into sexual behavior sharply exposes how, in turn, science is shaped by wider social ideas. Jordan-Young (2010) observes that certain behaviors were considered to be masculine or feminine in studies on sexuality, with libido, masturbation, and having multiple partners all considered part of masculine sexuality. However, in studies post–1980, these behaviors are considered to part of an acceptable definition of feminine sexuality (as long as the object-choice was male). As Jordon-Young (2010: 132) states, "As surprising as this is, what is *more* surprising is that the changes have gone unnoticed." Yet, of course, in the world beyond the laboratory, the sexual liberation of the 1960s and second-wave feminism had made their mark. Scientists do not conduct their research in a social vacuum.

Cordelia Fine further problematizes biological studies that emphasize differences in gender, to challenge what she calls "neurosexism." Like Jordan-Young, her work reviews hundreds of scientific papers and popular science books on the subject, carefully unpicking the data to reveal contradictions, generalizations, and hastily drawn conclusions. For example, on reviewing the impact of fetal testosterone, Fine observes a trend that emphasizes and reinforces the gender binary: experiments that report no measurable differences in relation to hormones are not deemed worthy of publication. She describes this as an example of "the so-called file-drawer phenomenon, whereby studies that *do* find sex differences get published, but those that don't languish unpublished and unseen in a researcher's file drawer" (Fine 2010: 134).

So gender difference is a self-perpetuating myth: one that is highlighted, magnified and, ultimately, distorted. But it also the lens of biologism, or rather, the prestige of science itself, that contributes to this distortion. As Fine (2010: 168) observes of neurology: "There's something special about neuroscientific information. It sounds so unassailable, so very ... well, *scientific,* that we privilege it over boring, old-fashioned behavioural evidence." So science itself is held in a hierarchy over other forms of understanding, apparently objective and therefore truthful, beyond criticism of the layperson. (To be fair, *Doctor Who* has taken advantage of our privileging of scientific knowledge, constructing technobabble to gloss over major plot points.)[5]

It is little wonder, then, that the 2018 series of *Doctor Who* glossed over the Doctor's change of sex. None of the companions, Yaz, Graham and Ryan, question the change on the few occasions where it was acknowledged. When the Doctor states in "The Woman Who Fell to Earth": "Half an hour ago I was a white-haired Scotsman. When's the next train due?" Ryan responds: "This is the last one back." Even a train timetable takes precedence over the switch in sex. When the Doctor praises the aforementioned Mehndi ritual in "Demons of the Punjab," Yaz prompts her to backtrack: "My references to body and gender regeneration are all in jest." Quite. So, while the Thirteenth Doctor provides a challenge to stereotypical perceptions of gender, namely through scientific abilities and competence (cf. Fine 2010; Orthia and Morgain 2016), the science of sex is carefully evaded.

Sex and Gender: A Reconciliation

In spite of the critique of the binary gender system by gender theorists, and those working in the sciences such as Fausto-Sterling, Jordan-Young and Fine, the perception of a rigid dichotomy prevails, as evidenced by Roberts' objections to trans rights and indeed the furor surrounding the Thirteenth Doctor. How is it possible to reconcile these two positions, or, at least, adopt an understanding that recognizes gender fluidity and adheres to scientific principles?

In 1985, biologist and feminist theorist Donna Haraway wrote her famous essay "A Cyborg Manifesto" which drew upon the concept of the cyborg as a metaphor to reconcile the divisions in feminism as it crossed the ruptures of race, class and geography in its attempts to speak for women. The cyborg, for Haraway, was a concept that broke down the divisions of animal and human, male and female, natural and unnatural: "a kind of disassembled and reassembled, postmodern collective and personal self. This is the self that feminists must code" (Haraway 1985: 163). While Haraway's essay is ironical in tone, freely describing the cyborg as both a fictional creature while also building an argument for intersectional politics, other authors have similarly sought to cross interdisciplinary boundaries when developing new approaches for thinking about the world. Philosopher Rosi Braidotti has argued for a paradigm shift away from the humanities' traditional concerns (which are generally based upon the work of white, male, middle-class scholars) to posthumanism, which itself challenges anthropocentrism:

> I refer to this move as expanding the notion of Life towards the non-human or *zoe*. This results in radical posthumanism as a position that transposes hybridity, nomadism, diasporas and creolization processes into means of re-grounding claims to subjectivity, connections and community among subjects of the human and non-human kind [Braidotti 2013: 50].

There is clearly a powerful conceptual, political potential for shifting our thinking beyond Western traditions and history. Obviously, the technological potential inherent in Haraway's cyborg has the ability to disrupt gender. After all, the Cybermen are seen to overwrite the sex of a would-be bride in "The Age of Steel" (2006), and on the two occasions where incarnations of Clara were imprisoned within a Dalek, the Dalek-machine spoke with a male voice ("Asylum of the Daleks," 2012; "The Witch's Familiar," 2015). However, as such technology is speculative—surgical reassignments aside—I am more interested in the potential we find in moving beyond our own species. After all, the binary of gender is an anthropocentric position. Turning to the wider animal kingdom, we swiftly find the binary of gender breaking down. Evolutionary biologist Peter Skelton has described just a few of the multiple ways that animals can not only change sex, but evade it altogether. For example, contrary to Hamer and Copeland's privileging of genetics, in some species, sex determination cannot be predicted by chromosomes. Male and female alligators have the same karyotypes, and so instead it is the incubation temperature of the egg that determines the sex of the hatchling (Skelton 1993: 89). Further still, sex differentiation is sometimes limited to certain generations. Aphids vary methods of reproduction according to the season of the year. In springtime, they reproduce parthenogenetically, resulting in an all-female generation. Later in the year, in autumn, they reproduce parthenogenetically again, with a new generation of males and females; this generation then reproduces sexually (Skelton 1993: 190–91).

Some species are sequential hermaphrodites, in that they change sex during their life cycle, either starting as male (protandrous) or starting as female (protogynous). The bluehead wrasse fish (*Thalassoma bifasciatum*) live in large female populations with just a couple of males. Removal of the males prompts several females to switch sex to replace them (Skelton 1993: 223–25), maintaining the sex ratio of the population. The phrase sequential hermaphrodite could reasonably be applied to the Time Lords, given both the Doctor and Master's change of sex; however the series has so far evaded examining the details of Time Lord reproduction, presumably to retain some mystique surrounding the species. Of course, one of the series' most enduring creations is the "hexapod hermaphrodite" Alpha

Centauri, first appearing in "The Curse of Peladon" (1972) and memorable enough to feature a cameo four decades later in "Empress of Mars" (2017). In "The Curse of Peladon," the Doctor relishes saying the phrase "hexapod hermaphrodite" to impress Jo; however, using the classifications above, and the suggestions in "The Timeless Children" (2020), the Doctor themselves may reasonably be defined as a "bipedal protogynous sequential hermaphrodite." Nevertheless the reproductive capabilities of Alpha Centauri are little more than a passing joke to highlight the alien nature of the character—much like the pregnancy of the Face of Boe (in "The Long Game," 2005), whom we later learn is Captain Jack Harkness (in "Last of the Time Lords," 2007). Conversely, gender fluidity isn't alien or inhuman. My point in evoking such examples in the animal kingdom is to problematize the idea that gender is somehow an inevitable result of biology. If "biology" is used as a synonym for "nature" and hence "natural," then we need to recognize that non-binary genders are indeed natural also. Taking a posthuman approach enables us to see that our distinctions between sex and gender are anthropocentric constructs and not universal conditions.

Conclusion

This essay has examined the Doctor's change of sex, or more particularly, approaches to sex and gender, and attempted to unpick the privileging of certain (but not all) scientific discourses that present gender as a rigid unalterable binary between male and female. Clearly, there are those who define themselves as cis-gendered men and women. My point is not to deny people their identities, nor the materiality of body, but rather the opposite—to emphasize the agency of the individual. Sex and gender are both complicated. The sexed body is result of multiple processes (sociological, medical, personal), and it does not necessarily match the gender identity of the individual.

At the start of this essay, I quoted Gareth Roberts' recourse to biology to uphold the rigidity of gender. However, it is clear that an examination of biology—whether genetics, endocrinology, or even body morphology—troubles this binary. It is possible to invert Roberts' own argument and say that it is indeed correct that no one is born into the wrong body. For example, a trans person was not born into the wrong body—it always has been their own body. Instead, it is the clumsy discourses of gender identification that failed to recognize their uniqueness at the outset; and they may shape and use their body as they please. (Somehow, I don't think this was what Roberts was intending.)

Scientific claims can be the result of the social beliefs at the time of the original research, and there is a sense that approaches to gender are shifting in response to society's awareness of gender complexity. Psychologist Christa Richards has argued that the medical profession needs to be more aware of gender fluidity in patient care, that genderqueer or non-binary identities are not considered part of any pathology (Richards et al. 2016). However, in this enlightened world, essentialist debates over causation remain. Reviewing gender diversity, Tinca J.C. Polderman and colleagues have sought to characterize gender identity as a "multifactorial complex trait with a heritable polygenic component" (Polderman et al. 2018: 95). This may be more complex than Hamer and Copeland's train tracks, but clearly genetic determinism continues to rear its head (although here emphasizing genotype over karyotype). Inevitably, we will continue to debate sex and gender for generations to come.

This is why the Thirteenth Doctor has enriched the series, bringing these issues to the fore. It is nevertheless a disappointment, however, that the Doctor remains white. Of course, scientific discourses of race are constructed in parallel processes to ones described above, and Jo Martin's surprise appearance in "Fugitive of the Judoon" raises equally exciting ideas that destabilize racial identity. But, to keep within the axes of gender, asking if the Doctor has changed sex or gender oversimplifies the question. The Doctor is genderfluid. The revelations in "The Timeless Children" suggest that we have simply failed to recognize the Doctor's femininity all along, and that the Thirteenth Doctor does not represent a radical departure from the Doctor's previous incarnations, but is part of a continuity that was always inherent in the character. Indeed, the change of sex in the latter episode seems far less significant than the revisions to series continuity concerning Gallifrey and the numbering of incarnations. So maybe, just maybe, our prescriptiveness to the gender system has changed, and, in some small way, we all regenerated ourselves along with the Thirteenth Doctor.

Notes

1. Roberts' argument is contradictory: in spite of claiming there is no such thing as gender identity, a dismissal of trans rights and activism serves to reify expectations that gender presentation should be congruent with sex as prescribed at birth. Rather than a challenge to gender identity (which he claims he doesn't believe), gender expectations are reinforced. Far from non-existent, gender identity is monolithic for Roberts.

2. We can speculate that there is an optimum height distance. For example, when Romana regenerates in "Destiny of the Daleks" (1979), the Doctor is compelled to reject her exceptionally short and tall options. Romana II, however, remains shorter than the Fourth Doctor in her final form.

3. The phallocentric nature of this train analogy is striking; I cannot help but speculate that the authors' descriptions of gender differences reflect their own perception of the sexes rather than an objective "scientific" description.

4. Gareth Roberts complains of that one of ills of trans awareness is that it may "medicalise children who don't conform to gender stereotypes" (Roberts 2019), yet the binary system of gender—of which he so disparagingly dismisses any violation—does precisely this.
5. Even this privileging of knowledge is gendered. Lindy Orthia and Rachel Morgain have used *Doctor Who* as a case study for the representation of scientists on screen. Although they report an increase in the frequency of depictions of female scientists across the decades, they note that scientific competence and independence are correlated with more stereotypical masculine characters, whereas those whose abilities are questionable are depicted as feminine (Orthia and Morgain 2016).

REFERENCES

Bigelow, S.J. (2013) Same old me, different face: Transition, regeneration, and change. In: Ellis, S., and Thomas, M.D. (eds.) (2013) *Queers Dig Time Lords: A Celebration of Doctor Who by the LGBTQ Fans Who Love It*. Des Moines, Iowa: Mad Norwegian Press, pp. 214–222.
Braidotti, R. (2013) *The Posthuman*. Cambridge: Wiley.
Butler, J. ([1990] 2008) *Gender Trouble: Feminism and the Subversion of Identity*. New York: Routledge.
Connell, R.W. (2009) *Gender*, 2nd edition. Cambridge: Polity Press.
Dickens, B.M. (2018) Management of Intersex Newborns: Legal and Ethical Developments. *International Journal of Gynecology and Obstetrics* 143: 255–259. doi:10.1002/ijgo.12573.
Fausto-Sterling, A. (1993) The Five Sexes, *The Sciences* 33 (2): 20–24. doi:10.1002/j.2326-1951.1993.tb03081.x.
Fausto-Sterling, A. (2000) *Sexing the Body: Gender Politics and the Construction of Sexuality*. New York: Basic Books.
Fine, C. (2010) *Delusions of Gender: How Our Minds, Society, and Neurosexism Create Difference*. London: Icon Books.
Flood, A. (2019, June 5) Doctor Who Anthology Drops Writer over Transgender Remarks. *The Guardian*. www.theguardian.com/books/2019/jun/05/doctor-who-anthology-writer-transgender-tweets-gareth-roberts.
Grosz, E. (1995) Sexual Difference and the Problem of Essentialism. In: *Space, Time and Perversion: Essays on the Politics of Bodies*. New York: Routledge, pp. 45–57.
Guardian Editorial (2017, July 17) *The Guardian* View on the Time Lord: Nurse Who? *The Guardian*. www.theguardian.com/commentisfree/2017/jul/17/the-guardian-view-on-the-time-lord-nurse-who.
Hamer, D.H., and Copeland, P. (1999). *Living with Our Genes: Why They Matter More Than You Think*. London: Macmillan.
Haraway, D.J. ([1985] 1991) A Cyborg Manifesto: Science, Technology, and Socialist-Feminisim in the Late Twentieth Century. In: *Simians, Cyborgs, and Women: The Reinvention of Nature*. London: Free Association Books, pp. 159–81.
Jordan-Young, R.M. (2010) *Brain Storm: The Flaws in the Science of Sex Differences*. Cambridge, Massachusetts: Harvard University Press.
Jowett, L. (2017) *Dancing with the Doctor: Dimensions of Gender in the Doctor Who Universe*. London: I.B. Tauris.
October, D. (2018) Hit or Miss? Fan Responses to the Regeneration of the Master. In: O'Day, A. (ed.) *Twelfth Night: Adventures in Time and Space with Peter Capaldi*. London: I.B. Tauris, pp. 235–59.
Orthia, L.A., and Morgain, R. (2016) The Gendered Culture of Scientific Competence: A Study of Scientist Characters in Doctor Who 1963–2013. *Sex Roles* 75 (3): 79–94. doi:10.1007/s11199-016-0597-y.
Polderman, T.J.C, Kreukels, B.P.C, Irwig, M.S., Beach, L., Chan Y-M, Derks, E.M., Esteva, I., Ehrenfeld, J., Den Heijer, M., Posthuma, D., Raynor, L., Tishelman, A., Davis, L.K., and International Gender Diversity Genomics Consortium (2018) The Biological

Contributions to Gender Identity and Gender Diversity: Bringing Data to the Table. *Behavior Genetics* 48 (2): 95–108. doi:10.1007/s10519-018-9889-z.

Richards, C., Bouman, W.P., Seal, L., John Barker, M., Nieder, T.O., and T'Sjoen, G. (2016) Non-Binary or Genderqueer Genders. *International Review of Psychiatry* 28 (1): 95–102. doi:10.3109/09540261.2015.1106446.

Roberts, G. (2019, 4 June) Statement on BBC Books and Transgenderism. *Medium.* medium.com/@zmangareth/statement-on-bbc-books-and-transgenderism-dd7ad0c9231a (accessed 5 November 2019).

Skelton, P.W. (1993) *Evolution: A Biological and Palaeontological Approach.* Harlow, England: Addison-Wesley.

West, C., and Zimmerman, D.H. (1987) Doing Gender. *Gender and Society* 1 (2): 125–51. doi: 10.1177/0891243287001002002.

YouTube (2017, July 16) Thirteenth Doctor Reveal. *Doctor Who.* youtu.be/_-_bSdWEYK8 (accessed 4 November 2019).

Candyfloss, Lego and Hope
What Sort of Scientist Is Jodie Whittaker's Doctor?

LINDY A. ORTHIA *and* VANESSA DE KAUWE

> THE DOCTOR: You're a medic. I'm the Doctor.
> MEDIC MABLI: A doctor of medicine?
> THE DOCTOR: Well, medicine, science, engineering, candyfloss, Lego, philosophy, music, problems, people, hope. Mostly hope.
> —"The Tsuranga Conundrum" (2018)

Introduction

The question of whether a woman could ever play *Doctor Who*'s lead character was debated by crew, commentators and fans for at least three and a half decades before Jodie Whittaker was cast as the Doctor in 2017 (Campion-Clarke 2017; Cordone and Cordone 2010). Script editor Christopher H. Bidmead floated the possibility when the BBC sought a replacement for Tom Baker in 1980, and Baker endorsed the idea, as did Patrick Troughton in 1983 (Campion-Clarke 2017; Capell 1983; Kingsley and Smylley 1980). Over the decades, the debate was often vitriolic, with some commentators employing sexist and homophobic arguments against a female Doctor (Campion-Clarke 2017). Even in 2019, the hashtag #NotMyDoctor retained currency among a subset of fans against it, despite Whittaker's first season shattering ratings records and her performance being widely praised (Belam and Martin 2018; Clarke 2018). It seems this, and related questions such as whether only white actors should be cast as the Doctor—notwithstanding Jo Martin's inspirational and ground-breaking portrayal of the character in 2020's "Fugitive of the Judoon" and "The Timeless Children"—are debates that will not fade easily (Hernandez 2013; Taylor 2017).

Diversity considerations, regarding the Doctor, matter for many reasons. However, a persistent thread of debate is the notion that the Doctor is an inspirational scientist role model, so casting a woman would challenge prevailing myths linking science careers to maleness. In 1986, a group of politicians touted this argument in a letter to the BBC, arguing a female Doctor would combat sexist perceptions of high tech jobs (*The Mirror* 1986). By the twenty-first century, the idea was more commonplace, with industry advocates asserting a female Doctor would raise the profile of women in science and counter the misconception that women are disinclined to study maths-intensive disciplines (Cook 2006; *The Telegraph* 2008). Ten years later, around the time Whittaker won the role, the idea had become mainstream, with popular culture commentators routinely reporting statistics about gender inequity in science and the desperate need for a female scientist role model of the Doctor's caliber (Deliee 2017; Mansuri 2017; White 2018).

Having now seen the first two seasons of the show under Whittaker's lead, it is timely to consider whether this campaigning paid off. What sort of scientist role model is Whittaker's Doctor? In this essay, we consider this question in three parts. First, we discuss whether she and previous Doctors can even be considered scientists.[1] We reflect on the Doctors' self-descriptors, plus definitions of "scientist" from history and sociology, and show that strictly speaking they don't fit the scientist label very well. Second, labels aside, we examine the scientific flavor of Whittaker's Doctor. As someone with vast knowledge and technical skills who engages in scientific-type activities, what ideological values does she embody with respect to science? Finally, we propose that she represents a version of expertise we call "scientific stewardship." This model emphasizes the ethical application of knowledge to benefit society, future generations and other species. We argue that in her peculiarly down-to-Earth version of this, Whittaker's Doctor sets a new benchmark for scientific role models in *Doctor Who*.

Is the Doctor a Scientist?

Self-Identification

Science commentators may think of the Doctor as a scientist, but if you ask the fourteen-plus Doctors themselves you'll get fifteen-plus different responses. Whittaker's Doctor is unique among new series Doctors in identifying as a scientist. The quote that opened this essay shows her claiming to be a doctor of medicine, science and engineering, bringing her close in her first season. Then, in "Praxeus" (2020), she describes herself and another character as "two brilliant scientists." The other new series Doctors talked around the issue: proclaiming their genius, allowing other characters to label them "scientist," and in David Tennant's case claiming to be a doctor "of everything." But none called themselves "scientist."

This is a world away from the classic series, in which William Hartnell's Doctor described himself as "a scientist," "an engineer," "a man of science" and "a doctor of science," and referred to his adventures as "scientific researches." All the classic series Doctors self-identified as scientists in different ways; all except Peter Davison and Sylvester McCoy, marking some movement away from identification with science in the 1980s (Orthia 2010). Overall, the early Doctors were much more obviously characterized as scientists via their labels, philosophies and actions than any Doctors appearing from 1980 onwards. It could be said that the early Doctors set a scientist tone explicitly, leaving the later Doctors room to play diverse variations on the theme.

Whittaker's embrace of the scientist label is thus only surprising if she is viewed within the scope of the new series. In this and other ways, her first two seasons represent a partial return to 1960s *Who* aesthetics and values. Indeed, her irreverent speech about being a doctor of "candyfloss," "Lego" and "hope" somewhat resembles an interaction between Troughton's Doctor and assassin Astrid in "The Enemy of the World" (1967–68):

> ASTRID: Oh, you're a doctor?
> THE DOCTOR: Well not of any medical significance.
> ASTRID: Doctor of law? Philosophy?
> THE DOCTOR: Which law? Whose philosophies, eh?
> ASTRID: I see you're determined to be mysterious.
> THE DOCTOR: Am I?
> ASTRID: Um, doctor of science? [...] A doctor of divinity then?
> THE DOCTOR: You'll run out of doctors in a minute.

Troughton's Doctor buries the nature of his expertise in playful mystery, while Whittaker's buries hers in outlandishness. But the effect is the same: dialogue associates both with science (and much more), without making any commitment to a specific discipline. The criterion of self-identification is thus inconsistent and vague when asking whether the Doctor is a scientist. As with so much else, the Doctor usually muddies the waters in this arena.

Rendering it still more unreliable, no Doctor has made any commitment to formal study or employment as a scientist. We know their academic performance on Gallifrey was poor compared to that of other Time Lords, including the Master and Romana. Their doctorate has always been dubious too: in "The Moonbase" (1967) they claimed to have taken a medical degree in Victorian Scotland, but by "The Ark in Space" (1975) said their doctorate is "purely honorary." The idea that they could have a job at all is mocked in "The Day of the Doctor" (2013) when they resume being scientific advisor for UNIT, after eschewing a UNIT salary in "Spearhead from Space" (1970), claiming to have "no use" for money. Alec Charles (2013: 95) argues the Doctor lacks a "sense of ongoing commitment"—a trait typical of a scientist—and asserts the Doctor's work is "a false and unfocused labour, their science a pseudo-science" and "they use the name of science as an excuse for their eclectic pursuits." The Doctor may enjoy being associated with science, but there is no discipline to it.

Historical Criteria

If the Doctor is not formally trained nor employed in science, can we consider them a scientist? In the twenty-first century Anglophone world, a scientist differs from other kinds of people participating in science by their qualifications and job, so the Doctor would likely not be accepted without question as part of today's scientist community. The professionalization of research is one of the traits historians recognize as constitutive of "science" in the strict sense: the knowledge system developed around 1800 CE that displaced natural philosophy and natural history in the West (Cunningham and Williams 1993). Until that time and even after it, few people had paid work as scientific researchers in Britain and similar countries, with research generally conducted by self-funded gentry or other kinds of amateurs (see Harmes and Scully, this volume). Meanwhile, in the world's thousands of other cultures throughout time, research into our bodies, the world and the universe has been organized in an infinite variety of ways, without the same kind of professionalization structure. "The scientist" is thus a very recent, culturally specific phenomenon. This is reflected in the English word "scientist," which was first printed in 1834 in a review of an acclaimed book by self-taught mathematician Mary Somerville (Anon. 1834; Somerville 1834). Somerville's book demonstrated connections between multiple sciences, thus prompting a search for a unifying term for the personnel.

This multi-disciplinary idea of a scientist fits the Doctor's mysterious and eclectic list of expert fields better than any discipline-specific term such as "physicist" or "biologist." However, the Doctor's additional expertise in matters we might consider *not*-science shows commonality with knowledge systems outside the modern Anglo tradition, that do not strictly demarcate between human and non-human phenomena. For example, the German concept of *Wissenschaft* unites the sciences, social sciences and humanities under an umbrella of systematic thought. In addition, many Indigenous Australian knowledge systems consider the ecological and social inseparable (Muir et al. 2010). Perhaps it is significant that in "State of Decay" (1980), the Doctor equated the word "scientist" with "witch wiggler," "wangateur," "fortune teller" and "mundunugu," all terms from knowledge systems outside of Western science that incorporate spiritual elements such as divination, magic or faith. This broader, more holistic view of expertise is in keeping with the Doctor's Time Lord nature: their practice and experience extends through all of time and space, and draws on the wide range of intellects preserved in the Time Lords' Matrix. It is also consistent with multi-faceted definitions of the title "doctor," which indicates healing as well as other kinds of expertise. This is reflected in Matt Smith's Doctor saying he is "here to help" ("Closing Time," 2011), with all the experiences his various lifetimes had brought. Thus in several ways, the Doctor does not neatly fit the scientist mold.

What then of the association between the title "Doctor" and research careers? The link between being a scientist and the title "doctor" is a relatively recent phenomenon. Universities first instituted the Ph.D. as a degree for research training in nineteenth century Germany, but British universities only adopted this model as recently as 1917 (Rospigliosi and Bourner 2019) and Australian universities even later. Even then, its application was haphazard, for example over half the academics recruited to British universities in the early 1960s did not have a doctorate (Rospigliosi and Bourner 2019). The clever but Ph.D.-less Doctor is thus true to the show's originating period and might have been a good fit for a 1960s British science department. Indeed, in "The Pilot" (2017), we discover the Doctor has been a lecturer at a Bristol university for possibly 70 years, and not in a conventional way, as Bill muses:

> BILL: I'm wondering what you're supposed to be lecturing on. It's like the university lets you do whatever you like. One time, you were going to give a lecture on quantum physics. You talked about poetry.

Once again this resonates with knowledge systems pre–1800 or beyond the West, and certainly not today's neoliberal universities. The titles "doctor" and "master" in fact originally stemmed from medieval European universities, where they functioned as terms of esteem for teachers rather than research qualifications (Rospigliosi and Bourner 2019). Arguably, this older "doctor/master" concept best characterizes what the Doctor does, and Ace's "Professor" nickname for McCoy's Doctor seems an apt variation under this model. To borrow from Kristine Larsen (2017: 102–103), the Doctor's approach to knowledge is not a single-minded interrogation of the universe driven by a search for useful applications, but rather "hearken[s] back to a medieval viewpoint of the interconnectedness of a nurturing cosmos that is knowable through respectful interactions with it." The Doctor employs their expansive expertise to assist, teach, inspire, reflect and communicate with others, much more than they engage in the original research expected of a scientist.

Sociological Norms

There is more to being a scientist than terminological definitions though. One well-known way of characterizing scientific work was proposed by American sociologist Robert Merton (1942) who described four institutional norms of the Western paradigm of science:

- *universalism*: knowledge claims are judged by impersonal criteria no matter who made the claim,
- *communalism*: knowledge is collaboratively produced and belongs to all,
- *disinterestedness*: research is done without motivations of personal gain, and
- *organized skepticism*: conclusions may be challenged with new information.

While some of these are no longer realistic norms of scientific work, they retain currency as ideals (Macfarlane and Cheng 2008). Arguably, the Doctor embodies them all at various points, though not consistently of course.

To begin with, any instance where the Doctor evaluates an idea on its merits could constitute universalism, and the program is littered with such interactions. Certainly, the Doctor seeks truth and is open to finding it in unusual places. However, they are susceptible to judging arguments by the speaker when they don't like them. For example, in "Arachnids in the UK" (2018), the obnoxious billionaire Robertson shoots a giant spider that is dying of suffocation and says it is a mercy killing. The Doctor objects, but merely asserts he is wrong rather than engaging with his argument, cruelly preferring to see the spider suffocate rather than give Robertson credit. More positively, the Doctor praises people they like for having useful ideas, especially when this is unexpected. While this is a delightful attitude in a scientist, it doesn't meet the measure of universalism because the Doctor's response is case-specific. Their tendency to judge people by ethical standards, to *see* individuals and tailor their interactions accordingly, is not consistent with dry, universalist objectivity. However, scientists for millennia have frequently failed to live up to this ideal themselves, with citation patterns and research trends remaining riddled with socio-culturally inflected bias (Harding 2006). So if the Doctor similarly fails, they have that in common with many real world scientists.

Regarding communalism, the Doctor is generally in favor of open knowledge and collaboration, at least in their overt rhetoric. The entire plots of serials like "The Krotons" (1968–69), "Full Circle" (1980) and "The Long Game" (2005) strongly advocate sharing knowledge for the public good. Yet there are moments when the Doctor implicitly endorses controlling information if it suits the goodies in a story, including in a subplot of "The Krotons" (de Kauwe and Orthia 2018). In addition they explicitly reject open knowledge at times. In "The Time Warrior" (1973–74), the Doctor advocates withholding technological knowledge from people he deems "uncivilized" by virtue of their historical period, class and/or culture. He thus breaches the communalist norm with questionable stagist reasoning (Orthia 2013b). He takes a similar stance in "The Long Game" when the unscrupulous Adam seeks to profit from knowledge of future technology. This exchange goes against the intention of the communalist norm, even if it is for understandable reasons.

Disinterestedness seems an easy fit for the anti-money, ethically upright Doctor, but once again, their strong ethical persuasion is incompatible with Merton's norms. Fundamentally, the Doctor uses science to right wrongs and prevent suffering and death—to sort out "fair play throughout the universe" ("The Woman Who Fell to Earth," 2018)—in all the variety of ways that may manifest, from racism to exploding spaceships. The Doctor is opinionated, their actions guided by values as well as curiosity. Thus, their actions are frequently based on *interested* motivations, rather than *disinterested* ones. Admittedly, this is one reason Merton's norms have been questioned by scientists in recent decades, who similarly want to pursue research they believe will do good in the world (Macfarlane and Cheng 2008). So while the Doctor may not meet this criterion, they are not alone.

Finally, organized skepticism is partly a soft version of the scientific method, in which the best hypothesis is proposed then tested as new information becomes available. Some Doctors have repeatedly breached this norm by arrogantly asserting knowledge claims that cannot be questioned because of their godlike perspective, a tendency founding producer Verity Lambert described as "this awful thing of knowing everything and being right about everything" (Tulloch and Alvarado 1983: 130). She said this specifically about Tom Baker's Doctor, but Jon Pertwee's and Tennant's Doctors were similarly godlike (Orthia 2010). As many commentators have pointed out, Whittaker's Doctor is not like that, instead regularly acknowledging when her information is incomplete and when she has made a mistake. Some critics have decried her lack of godlike omniscience compared to other new series Doctors (e.g., see comments by JamesValencia, Furry-Canary and topica at Belam and Martin 2018), while others have praised the fact that she is "less of an insufferable know-it-all than almost any previous Doctor" and "does not (as most previous Doctors have done) bully with her intellect or authority" (McDunnah 2019). In this, she embodies the norm of organized skepticism and demonstrates good scientific practice by normative standards.

All things considered, the Doctor does not fully embody the role of scientist by Western sociological or historical definitions. At the same time, they and *Doctor Who* are so inflected with Britishness (Nicol 2018a), it would be difficult to make a case that the Doctor embodies an equivalent role in a non–British culture. But they may yet embody some other form of knowledge keeper, which takes into account the Time Lord's embrace of all of time and space. It may be that the TARDIS translation circuit translates this role as "scientist" because it is the closest English equivalent, even if it is inadequate (cf. Bowker and Halley this volume).

If so, what is sciency about the Doctor?

What Scientific Traits Does the Doctor Possess?

Doing Science

If self-identification, employment conditions and sociological norms are unlikely to characterize most Doctors as scientists, the question remains why some viewers continue to associate them with that role. The most obvious reason is their expansive knowledge and technical skill. All incarnations of the Doctor have possessed these traits, even if some revealed knowledge gaps and fallible skills. The Doctor knows things that astronomers, physicists, biologists, chemists and geologists would like to know, as well as reiterating knowledge scientists in those disciplines already know. They are also able to fix and heal, build and redesign things that engineers, computer scientists and medical professionals would like to be able to fix, heal, build and redesign. Having abundant knowledge and skills in these fields marks the Doctor as a scientist in the sense of being someone who engages in work considered "scientific" in the Western imaginary.

Whittaker's Doctor certainly engages in a large amount of this knowing and fixing during her first two seasons. Most of her knowledge is alien in nature: knowledge of aliens' biology, cultures and planets, and alien technologies such as ships, weapons and energy sources, robots, biotech and bodily implants. But some is Earth-level science, such as the properties of acetylene, environmental conditions that attract or repel spiders, the nature of antimatter, and the structure of computer software, not to mention the consequences of climate change and plastic pollution. Meanwhile her "serious tech skillz" ("Resolution," 2019) include forging a sonic screwdriver from Sheffield steel cutlery, creating a time/space transport device from a microwave oven and car battery, analyzing alien dust using materials from a 1940s Indian/Pakistani homestead, and checking alien proteins with chicken eggs. Throughout, she is led by relentless optimism—hope—as well as boundless curiosity and an inquiring mind that never ceases to investigate until she has found the solution to a problem or answered a question. Like other Doctors, she is a creative thinker who cobbles her prior knowledge, new information and available materials together to discover, unravel and create phenomena she hasn't encountered before. There is indeed something very sciency about all this.

The leap to then label the Doctor a "scientist" stems from *Doctor Who*'s genre, science fiction. Science fiction is less interested in exploring the mundane contexts of present-day scientific labor; more interested in criticizing Western science's ethical excesses or promoting its value as a truth-finding method (Haynes 1994; Weingart et al. 2003). The scientist figure thus has a symbolic function in much science fiction rather than a representational one: to symbolize science's ethical-political milieu, not to represent the scientist's life. Sci-fi scientist characters are often identifiable solely because of their knowledge and skills, not because of their employment conditions, and the Doctor shares this trait. So what symbolic values does the scientist Doctor stand for?

Ideologies of Science

The answer is complex: each Doctor has symbolized a different set of ideological perspectives re science. For example, Pertwee's Doctor frequently embodied a scientistic ideology, as did Tom Baker's Doctor during his gothic horror years, whereas Troughton's tenure was marked by ideological ambivalence about science and critique of it (Orthia 2010, 2011). The Doctor's symbolic function in fact modulates with respect to the scientific tenor of the companions (Orthia 2010; Tulloch and Alvarado 1983). Pairing the Doctor with "clothhead" Jo or "savage" Leela promoted an authoritarian version of scientist-as-truth-knower, whereas pairing him with both "genius" Zoe and less educated Jamie allowed him to embody a more nuanced, fallible version of scientist. The ideological differences are neatly summarized in key quotes from "The Dæmons" (1971), "The Robots of Death" (1977) and "The Wheel in Space" (1968) respectively:

> PERTWEE: Everything that's happened in life must have a scientific explanation.
> BAKER: To the rational mind nothing is inexplicable, only unexplained.
> TROUGHTON: Logic, my dear Zoe, merely enables one to be wrong with authority.

As with the Troughton era, the Hartnell and Davison eras were marked by ideological ambivalence regarding science. This was again partly because of the ensemble TARDIS crew that comprised at least one other scientifically or intellectually expert character, eliminating the need for the Doctor to carry the burden of representing science solo. These eras showed there are different ways to be a scientist, thus rendering the scientist role a legitimate target for debate.

The situation is somewhat similar for Whittaker's Doctor. She is not godlike; she makes mistakes and asks questions of others to fill gaps in her knowledge. In this she resembles Hartnell, Troughton and Davison's Doctors, all of whom exhibited these traits repeatedly (Orthia 2010). They were also the only Doctors to pursue purely historical adventures (stories with no science fiction element). While the same is not quite true of Whittaker, "Demons of the Punjab" (2018) comes close in that the Doctor, her companions and the visiting aliens are mere witnesses to a violent past, not significant actors in it (Nicol 2019). "The Ghost Monument" (2018) also shares much with early *Doctor Who* stories, including "An Unearthly Child" (1963), whose plot primarily focused on the crew trying to get back to the TARDIS to escape a bad situation. Such plots characterize Whittaker's Doctor as somewhat powerless in the face of unalterable time and vast space. Like Hartnell, Troughton and Davison, she is humbled by it, not omnipotent. The Timeless Child arc unifying her second season reinforces this still further, as she discovers significant aspects of her own life that she does not remember at all, let alone have command of. This contrasts with other Doctors such as Tennant's, who was authoritative about history's fixed points, and memorably abused his Time Lord power to challenge this inalterability in "The Waters of Mars" (2009).

A Hartnell, Troughton and Davison aesthetic is also present in the number of Whittaker's traveling companions. As with the earlier Doctors, this allows multiple voices and perspectives to be heard in her stories. Arguably, the perspectives of Ryan, Yaz and Graham are vastly more diverse than ever before given their unprecedented diversity in class, ethnicity, region, religion, (dis)ability and gender balance. However, an important difference is that none are scientifically trained. This leaves Whittaker's Doctor as science's sole representative in her TARDIS, so there are no debates about how a scientist should behave in her first two seasons. In fact, this Doctor's companions do not argue with her at all; unusually, the dynamic is one of harmonious teamwork, the Doctor referring to her companions as her "fam," and the companions singing her greatness. In fact, the fam only argue with her when she's endangering her own existence. As a result, the companion-Doctor relationship is not a place where conflicts over good and evil—or over science—take place.

Further, in Whittaker's first season, the locus of ethical debate is largely located outside science in keeping with new series trends. There are two exceptions, but both ultimately point away from science when laying blame for disaster. "The Battle of Ranskoor Av Kolos" (2018) resembles a mad-scientist story in that vengeful alien Tzim-Sha captures and shrinks whole planets by combining his own species' powerful technology with another species' telekinetic abilities. In doing so, he risks tearing open the fabric of the universe, but science and technology are not blamed in the story; rather, it is Tzim-Sha's power-hungry brutality that is subject to criticism. Similarly, in "Arachnids in the UK," a giant spider plague is blamed on the negligent billionaire who dumped biological waste in a coalmine, upon which a mutant spider fed and grew. Somehow, the genetic scientists who created and accidentally disposed of the mutant spider are relieved of responsibility in the story, since the Doctor never blames them. The unintended downstream applications of science are criticized, but not scientific research and products themselves. Other scientist characters in this season are either victims (e.g., the dead, enslaved scientists in "The Ghost Monument") or benignly helpful (e.g., medic Mabli and engineer Durkas in "The Tsuranga Conundrum").

This flavor changes slightly in Whittaker's second season, to a more active endorsement of some aspects of science and scientists. For example, in "Orphan 55" (2020), on a future Earth devastated by climate change, the Doctor tells her companions rather passionately, "You had warnings from every scientist alive." Even in "Praxeus," while criticizing as unethical a group of alien scientists who used Earth as an experimental medical lab, the Doctor quickly passes over this and tries to help them complete their work. She also champions the philosophy of Islamic physicians when arriving in 1380 CE Syria, for their pioneering hospital and progressive treatment of people with mental health problems ("Can You Hear Me?," 2020). She admires the contributions to science and technology of figures she meets, including Ada Lovelace, Charles Babbage and Nikola Tesla, but also non-scientist historical figures like the pacifist spy Noor Inayat Khan. So, while giving a quiet thumbs up to some scientists, she does not promote Western science as a belief system over and above other values in the way Pertwee's or Tom Baker's Doctors did. Thus, Whittaker's Doctor does not symbolize "hero scientist" or "evil scientist" as such, since science is relatively incidental, not core, to her stories' moral messages. The ideological flavor of her era re science is not ambivalence; it is more of an understated admiration in concert with admiration for other ways of approaching the universe. Science is a vehicle to her, not an end in itself.

Given this, how may we characterize this Doctor's role more precisely? Who is this person who engages in scientific knowing and doing, yet does not strongly represent criticism or advocacy of science, and is not a scientist?

Whittaker's Doctor as Science Steward

The Doctor is primarily a collector, user and communicator of knowledge, more than they are a researcher as such. They are also an ethical actor whose primary motivation is to do good. This combination of traits suggests the notion of stewardship. Worrell and Appleby (2000: 263) define stewardship as "the responsible use […] of natural resources in a way that takes full and balanced account of the interests of society, future generations, and other species, as well as of private needs." It is not just about natural resources though; it is equally possible to refer to the stewardship of knowledge and practical skills (Ross et al. 2016; Saltman and Ferroussier-Davis 2000). Based on this, we suggest "science stewardship" might be defined as the responsible use, nurture and development of various knowledge systems in a way that takes into account the interests of society, future generations, and other species, as well as of private needs. This sounds a lot like what the Doctor strives for.

To posit that Whittaker's Doctor is a steward of science accounts for both her lack of allegiance to any specific knowledge system and her extensive knowledge and technical skills. She is not technically a scientist, nor is she scientistic. Rather, she is a steward of knowledge sourced from across time and space. This collation of learning helps her discern and respond to each new situation appropriately. This does not discount Nicol's (2018a) assertion of the Doctor's quintessential Britishness, rather it confirms the broader context of his assertion: that *Doctor Who* is both a reflection of Britain and *a projection of progressive aspirations for it*, including its multicultural potential. Thus, *Doctor Who* frequently reflects the ideological needs of the current climate. This may include the need to break from traditional Western models for managing knowledge and embrace more holistic stewardship of knowledge, science and technology (Saltman and Ferroussier-Davis 2000).

In this stewardship capacity, it is conceivable that Whittaker's Doctor might, in future seasons, engage meaningfully with real world human knowledge systems beyond Western science. For example, as a Time Lord with all of time and space at her disposal, it should come as no surprise to see the Doctor repeatedly visit some of the 250+ Indigenous Nations from across the Australian continent, custodians of the oldest continuing cultures in the world, with robust knowledges and modes of science refined through tens of millennia. Parts of Whittaker's first two seasons, especially "Rosa" (2018), represented race topographies with great nuance compared to previous eras of *Doctor Who* (Gooden 2018), so perhaps this is the first time the program can be trusted to engage with Earth's First Peoples' knowledges respectfully. Though that's a big "perhaps," given *Doctor*

Who's past depictions of Indigenous peoples (McMurtry 2013; Morgain 2013; Orthia 2013a), so vigilance will always be required.

The term "stewardship" can carry religious connotations, and it is true that Judeo-Christian and other world religions consider stewardship an important concept. But scholars generally concur it is about living well with each other and with other forms of life (Tickell 2006: 223), and that it neither excludes nor necessitates a religious element (Berry 2006). Accordingly, Whittaker's accessible and down-to-earth Doctor is with and for the everyday person, as well as other beings she may encounter. But this does not mean she is presented in a divine or even pious persona. Rather, like previous Doctors, she knows she is clever and is not ashamed of it, proclaiming at various points, "of course it worked, I'm not an amateur," "I'm good at building things, probably," and "did I not mention? I am really smart." This at times means other characters including companions remain baffled by her words and actions. More than that, some scripts have her hogging the intellectual limelight where her companions are concerned. They are given some opportunities to shine, but the Doctor on occasion steals their thunder: trumping nurse Grace's medical knowledge in "The Woman Who Fell to Earth," and in "Resolution" entirely ignoring Yaz's status as a police officer when using CCTV, GPS and number plate tracking to follow a Dalek disguised as a cop. So while in "The Witchfinders" (2018) Graham asserts the group has "a very flat team structure, we all have our area of expertise," and the Doctor later echoes this twice, the structure as written is in fact not really flat. A character in "Fugitive of the Judoon" remarks on this, telling the Doctor, "You're the smartest. I can see it in your eyes." More than this, in "The Haunting of Villa Diodati" (2020), the Doctor declares, "'Cause sometimes this team structure isn't flat. It's mountainous, with me at the summit in the stratosphere, alone, left to choose."

In this, Whittaker's two-season launch fails to live up to its potential in truly democratizing expertise, even while it dangles the possibility before us. The stewardship model the Doctor seems to embody is not inherently democratizing, since participants need to be inducted into respectful and ethical norms before given power (Ross et al. 2016), but it does invite participation. Unlike Western science, it does not police the boundaries of truth or assert that truth is singular. Whittaker's Doctor seems to share these values, even if she is sometimes scripted as overriding them.

That there is a new *Doctor Who* philosophy on scientific and technological expertise is well illustrated in the backyard-tinkerer style of the Doctor's engineering and applied science. Her welding-mask-and-ox-spit mechanics are decidedly down to Earth, in a departure from new series bing-when-there's-stuff timey-wimeyness. The Doctor's technology, while undoubtedly speculative, *feels* accessible. Her expertise seems earned—built—rather than transcendent. The only differences between the Doctor and us are her longevity and TARDIS that mean she has "lived longer, seen more, loved more and lost more" than the rest of us ("It Takes You Away," 2018), and perhaps also her alien culture that renders her socially awkward with human families (though for many of us that isn't so different). There can be no coincidence between this and the presence of not one but two unequivocally working class companions in Ryan and Graham, unprecedented in the program (Nicol 2018b). Whittaker is far from the mysticism of Tennant's "lonely god," nothing like the roller-coaster weirdness of Smith's "question hidden in plain sight." As she says in her first episode, "I'm just a traveller. Sometimes I see things need fixing, I do what I can." More than perhaps any other Doctor, Whittaker presents science and technology as ordinary, everyday, relatable: potentially belonging to all of us.

Conclusion

To return to our opening theme, this active, skilled and knowledgeable persona sets the Doctor apart from many fictional female scientists. The history of film is riddled with female scientist characters who surrender their intelligence for love, are book-smart yet field-incompetent, or are assistants to senior male scientists (Flicker 2003). Female scientists with proven intelligence are often evil or negatively depicted as asexual or masculine (Flicker 2003). Whittaker's Doctor is far from any of these stereotypes. She maintains authority, intelligence and skills throughout her season, and is neither gendered nor sexualized in a negative way by the program. Consistent with *Doctor Who*'s record in depicting non-regular female scientists (Larsen 2018; Orthia and Morgain 2016), but a big improvement on depictions of female scientist companions (Jowett 2014; Orthia 2010), she is just as scientifically credible as her male incarnations. As such, she lives up to advocates' hopes for a prominent female role model. She demonstrates repeatedly, and without fanfare, that women can wield scientific knowledge and engage in scientific activities with competence, ease and authority. In this and in her steward persona, Whittaker's Doctor may not be a scientist, but she is charting new territory as a role model in science.

Notes

1. We consider only the televised serials, not stories in other media or spinoffs.

References

Anon. (1834) Art. III.—*On the Connection of the Physical Sciences. By Mrs. Somerville. Quarterly Review* 51: 54–68.
Belam, M., and Martin, D. (2018, December 11) "Too touchy-feely"? Our panel on Jodie Whittaker's first series of Doctor Who. *The Guardian.* www.theguardian.com/tv-and-radio/2018/dec/10/too-touchy-feely-our-panel-on-jodie-whittakers-first-series-of-doctor-who.
Berry, R.J. (2006) Introduction. In: Berry, R.J. (ed.) *Environmental Stewardship.* London: T&T Clarke, pp. 1–13.
Campion-Clarke, E. (2017, July 29) The signs were there. *The True Evil of Ewen Campion-Clarke.* dempsyobrien.blogspot.com/2017/07/the-signs-were-there.html (accessed 16 October 2019).
Capell, L. (1983) Episode dated 17 March 1983. *Nationwide.* UK: BBC.
Charles, A. (2013) Three characters in search of an archetype: Aspects of the trickster and the *flâneur* in the characterizations of Sherlock Holmes, Gregory House and Doctor Who. *Journal of Popular Television* 1(1): 83–102. doi:10.1386/jptv.1.1.83_1.
Clarke, S. (2018, December 10) Jodie Whittaker to return as "Doctor Who" in 2020 amid strong U.S. ratings. *Variety.* variety.com/2018/tv/news/doctor-who-jodie-whittaker-return-2020-strong-us-ratings-1203085568/ (accessed 16 October 2019).
Cook, M. (2006, August 14) It's science, Jim, but not as we know it. *The Age.* www.theage.com.au/news/education-news/its-science-jim-but-not-as-we-know-it/2006/08/12/1154803138238.html.
Cordone, J., and Cordone, M. (2010) Who is the Doctor? The meta-narrative of *Doctor Who.* In: Hansen, C.J. (ed.) *Ruminations, Peregrinations, and Regenerations: A Critical Approach to Doctor Who.* Newcastle upon Tyne: Cambridge Scholars Publishing, pp. 10–21.
Cunningham, A., and Williams, P. (1993) De-centering the "big picture": *The Origins of Modern Science* and the modern origins of science. *The British Journal for the History of Science* 26(4): 407–432. doi:10.1017/S0007087400031447.
de Kauwe, V., and Orthia, L.A. (2018) Knowledge, power and the ethics illusion: Explaining diverse viewer interpretations of the politics in classic era *Doctor Who. Journal of Popular Television* 6(2): 151–165. doi:10.1386/jptv.6.2.151_1.
Deliee, M. (2017, July 16) It's time for a black female Time Lord on Doctor Who. *PopSugar.* www.popsugar.com/news/Who-Next-Doctor-Who-Actor-43696898 (accessed 30 September 2019).
Flicker, E. (2003) Between brains and breasts—women scientists in fiction film: On the marginalization and sexualization of scientific competence. *Public Understanding of Science* 12: 307–318. doi:10.1177/0963662503123009.
Gooden, T. (2018, October 23) "Doctor Who" 11x03 review: "Rosa" sets a new standard for the sci-fi series. *Hypable.* www.hypable.com/doctor-who-11x03-rosa-review/ (accessed 18 November 2019).
Harding, S. (2006) *Science and Social Inequality: Feminist and Postcolonial Issues.* Urbana and Chicago: University of Illinois Press.
Haynes, R.D. (1994) *From Faust to Strangelove: Representations of the Scientist in Western Literature.* Baltimore: John Hopkins University Press.
Hernandez, M. (2013) "You can't just change what, I. look like without consulting me!": The shifting racial identity of the Doctor. In: Orthia, L. (ed.) *Doctor Who and Race.* Bristol: Intellect, pp. 45–60.
Jowett, L. (2014) The girls who waited? Female companions and gender in *Doctor Who. Critical Studies in Television* 9(1): 77–94. doi:10.7227/CST.9.1.6.
Kingsley, H., and Smylley, P. (1980, October 25) Tom Baker quits after seven years. *The Mirror.* cuttingsarchive.org/images/e/e5/1980-10-25_Daily_Mirror.jpg.
Larsen, K. (2017) Medieval organicism or modern feminist science? Bombadil, elves, and Mother Nature. In: Vaccaro, C., and Kisor, Y. (eds.) *Tolkien and Alterity.* Switzerland: Palgrave, pp. 95–108.
Larsen, K. (2018) The river, the rock, the relative and the returned: Depictions of women scientists in *Doctor Who*'s Moffat era. In: Carlson, A.L. (ed.) *Women in STEM on Television.* Jefferson: McFarland, pp. 187–206.

Macfarlane, B., and Cheng, M. (2008) Communism, universalism and disinterestedness: Re-examining contemporary support among academics for Merton's scientific norms. *Journal of Academic Ethics* 6: 67–78. doi:10.1007/s10805-008-9055-y.

Mansuri, N. (2017, July 19) "Doctor Who" season 11: What we want to see now that the Doctor is a woman. *Hypable*. www.hypable.com/the-doctor-is-a-woman/ (accessed 30 September 2019).

McDunnah, M.G. (2019, January 16) Doctor He, She, or They? Changing gender, and language, in *Doctor Who*. *Conscious Style Guide*. https://consciousstyleguide.com/doctor-he-she-or-they-changing-gender-and-language-in-doctor-who/ (accessed 24 October 2019).

McMurtry, L. (2013) Inventing America: *The Aztecs* in context. In: Orthia, L. (ed.) *Doctor Who and Race*. Bristol: Intellect, pp. 103–107.

Merton, R.K. (1942) A note on science and democracy. *Journal of Legal and Political Sociology* 1: 115–126.

The Mirror (1986, December 20) Sex change for Dr Who! cuttingsarchive.org/images/9/90/1986-12-20_Daily_Mirror.jpg.

Morgain, R. (2013) Mapping the boundaries of race in *The Hungry Earth/Cold Blood*. In: Orthia, L. (ed.) *Doctor Who and Race*. Bristol: Intellect, pp. 251–267.

Muir, C., Rose, D., and Sullivan, P. (2010) From the other side of the knowledge frontier: Indigenous knowledge, social-ecological relationships and new perspectives. *The Rangeland Journal* 32: 259–265. doi:10.1071/RJ10014.

Nicol, D. (2018a) *Doctor Who: A British Alien?* Switzerland: Palgrave Macmillan.

Nicol, D. (2018b, October 16) When you're in space the whole cosmos is British. *Politics and Law of Doctor Who*. politicsandlawofdoctorwho.blogspot.com/2018/10/when-youre-in-space-whole-cosmos-is.html (accessed 29 October 2019).

Nicol, D. (2019, October 24) From Paris to the Punjab: Religious intolerance and the time traveller's plight in BBC's Doctor Who. *Woolf Institute Blog*. www.woolf.cam.ac.uk/blog/from-paris-to-the-punjab-religious-intolerance-and-the-time-travellers-plight-in-bbcs-doctor-who (accessed 28 October 2019).

Orthia, L.A. (2010) *Enlightenment was the Choice: Doctor Who and the Democratisation of Science*. Ph.D. Thesis, The Australian National University.

Orthia, L.A. (2011) Antirationalist critique or fifth column of scientism? Challenges from *Doctor Who* to the mad scientist trope. *Public Understanding of Science* 20(4): 525–542. doi:10.1177/0963662509355899.

Orthia, L.A. (2013a, May 23) A very good googly—race in Four to Doomsday. *Doctor Who and Race*. doctorwhoandrace.com/2013/05/23/a-very-good-googly-race-in-four-to-doomsday/ (accessed 18 November 2019).

Orthia, L.A. (2013b) Savages, science, stagism and the naturalized ascendancy of the Not-We in *Doctor Who*. In: Orthia, L. (ed.) *Doctor Who and Race*. Bristol: Intellect, pp. 269–287.

Orthia, L.A., and Morgain, R. (2016) The gendered culture of scientific competence: A study of scientist characters in *Doctor Who* 1963–2013. *Sex Roles* 75: 79–94. doi:10.1007/s11199-016-0597-y.

Rospigliosi, A., and Bourner, T. (2019) Researcher development in universities: Origins and historical context. *London Review of Education* 17(2): 206–222. doi:10.18546/LRE.17.2.08.

Ross, A., Pickering Sherman, K., Snodgrass, J.G., Delcore, H.D., and Sherman, R. (2016) *Indigenous Peoples and the Collaborative Stewardship of Nature*. London: Routledge.

Saltman, R.B., and Ferroussier-Davis, O. (2000) The concept of stewardship in health policy. *Bulletin of the World Health Organization* 78(6): 732–739.

Somerville, M. (1834) *On the Connexion of the Physical Sciences*. London: John Murray.

Taylor, C. (2017, July 21) A black "Doctor Who" is long overdue too, says former Doctor. *Mashable*. mashable.com/2017/07/21/black-doctor-who/ (accessed 16 October 2019).

The Telegraph (2008, December 1) "Doctor Who should be a woman" say female scientists. www.telegraph.co.uk/news/celebritynews/3538551/Doctor-Who-should-be-a-woman-say-female-scientists.html.

Tickell, C. (2006) Religion and the environment. In: Berry, R.J. (ed.) *Environmental Stewardship*. London: T&T Clark, pp.220–227.

Tulloch, J., and Alvarado, M. (1983) *Doctor Who: The Unfolding Text*. London: Macmillan.

Weingart, P., Muhl, C., and Pansegrau, P. (2003) Of power maniacs and unethical geniuses:

Science and scientists in fiction film. *Public Understanding of Science* 12: 279–287. doi:10.1177/0963662503123006.

White, M. (2018, October 8) The new Doctor Who picks up the chase with a pace as she crosses the gender barrier. *The Conversation.* theconversation.com/the-new-doctor-who-picks-up-the-chase-with-a-pace-as-she-crosses-the-gender-barrier-104312.

Worrell, R., and Appleby, M.C. (2000) Stewardship of natural resources: Definition, ethical and practical aspects. *Journal of Agriculture and Environmental Ethics* 12(3): 263–277. doi:10.1023/A:1009534214698.

The Mad Scientist Wore Prada
Female Frankensteins in the Universe of Doctor Who

KRISTINE LARSEN

Introduction: Scientific Villains and Vixens

Western science can seem to be primarily the domain of men. While female scientists may no longer be odd exceptions, the statistics suggest that women in the physical sciences are far from reaching parity with men. In 2016 women earned only 18 percent of the Ph.D.s in physics awarded in the USA, down from a high of 23 percent in 2000 and 2003 (American Institute of Physics n.d.). Numerous authors have made the case that continued media depictions of science and scientists that portray science, technology, engineering, and mathematics (STEM) as inherently masculine fields can have a chilling effect on increased female participation (Chimba and Kitzinger 2010; Geena Davis Institute 2018; Haran et al. 2008; Saucerman and Vasquez 2014).

Doctor Who has been "rich in science content" over its five-plus decades, making it a logical sandbox in which to study depictions of scientists (Orthia 2019: 3), including negative stereotypes about gender. Roslynn Haynes (2003: 244) identified seven common stereotypes in depictions of scientists in Western literature—the Faustian "evil alchemist"; the absent-minded "foolish scientist"; the Frankensteinian "inhuman researcher"; the "mad, bad, dangerous scientist" perhaps best represented by H.G. Wells' Doctor Moreau; the "helpless scientist" whose work cannot be controlled; the Indiana Jones–like "scientist as adventurer"; and the heroic "noble scientist." Only the last two have positive attributes. In her 2010 Ph.D. thesis *Enlightenment Was the Choice: Doctor Who and the Democratisation of Science*, Lindy Orthia specifically analyzed non-recurring (not companion) scientists in episodes broadcast between 1963 and 2008 and found no examples of Haynes' "scientific adventurer" type. She instead grouped all ethical scientists into the "noble scientist" category and added a new category, "scientist victim," those who are "not free to make ethical choices," usually due to the control of others (198). In her 2003 study of feature films, Eva Flicker argued that the "cliché description of 'mad scientist' does not apply to women scientists. They do not work in hidden laboratories on dubious projects but rather, remain solid 'with their feet on the ground'" (316). While this may have been true of the sample she selected, the fact that a parody of a female mad scientist—the murderous Pearl Forrester—was central to the cable television comedy series *Mystery Science Theater 3000* (1988–99, 2017–18) for eight years from 1992 to 1999 suggests that this is not broadly true. Other studies have indeed found that the percentage of female scientists depicted as "mad" or "inhumane" is significantly lower than their male counterparts (Geena Davis Institute 2018; Steinke 2005), but such examples do exist. In her study Orthia (2010: 196–97) found that only 21.5 percent of non-recurring scientist characters are female and there are far more female "noble scientists" than female "mad scientists" or female "inhumane researchers" (making up 54.3 percent as opposed to 8.6 percent and 20 percent, respectively), but they weren't all noble.

The fact that fewer fictional female scientists are negatively depicted as "mad" does send a potentially significant message to consumers of popular media in terms of gender and ethics; however, there is a less positive implication looming just beneath the surface. Villains are often the most complex (and hence memorable) characters in a work of fiction, and along with heroes and generic protagonists represent "the most prominent and therefore the most influential" characters (Geena Davis Institute 2018: 11). If women are depicted as less likely to be meaningful villains, there is a danger of sending the message that their stories are less worth telling. Conversely, it can draw additional attention to those few examples in unhelpful ways, given research demonstrates that when women scientists—either fictional examples or real-world scientists—are depicted in media there is an unbalanced emphasis placed upon their appearance, and in some cases their romantic partners and sexuality (Chimba and Kitzinger 2010; Haran et al. 2008; Steinke 2005).

This essay will deconstruct the depictions of three women in the Whoverse who fit the definition of a mad scientist, inhumane researcher, or both—the Rani, Madame Kovarian, and Missy—with the intention of identifying negative stereotypes and identifying the mixed messages their portrayals send to the audience.

Depictions of Scientists in Doctor Who

Jess Nevins (2011) argued that female mad scientists can be found as early as Alexander Pope's satirical poem "The Dunciad" (1728), where scientific knowledge is personified as "Mad Mathesis.... Too mad for mere material chains to bind." Nevins identified three female mad scientists in late nineteenth century novels: the titular characters of George Griffith's *Olga Romanoff* (1893–94) and T. Mullett Ellis' *Zalma* (1895), and Madame Koluchy in Robert Eustace's *The Brotherhood of the Seven Kings* (1898–99). Unlike their male counterparts, these female mad scientists (who create weapons of mass destruction) were "primarily active outside their laboratories," and "portrayed as sexual beings, either using their sexual attractiveness to manipulate men or being sexually profligate as a sign of their moral perversity [...] Male mad scientists were usually passionless (though not emotionless), where female mad scientists were passionate" (Nevins 2011). Flicker (2003: 316) argued that this emphasis on female scientists' passion (in both a scientific and sexual sense) adds "intuition, emotional elements, love affairs, and feelings" to the narrative, but because of this, women scientists "do not represent the rational scientific system of their male colleagues. They are therefore taken less seriously as 'scientists.'" For example, Nicholes (2012: 125) found that the television series *Blindspot* (2015–present) depicts women who "mismanage their emotions and reinforce harmful stereotypes [...] The female characters either try to ignore their emotions or are completely incapacitated by them." Robotic hyper-rational stereotypes are also seen in depictions of female forensic scientists (Haran et al. 2008: 25). Coupled with representations of women scientists as antisocial or quirky, such depictions paint women scientists as "not characters with whom average viewers can relate" (Nicholes 2012: 120).

Orthia documented that within the world of *Doctor Who* insanity is literally equated with "unreason"; therefore mad scientists are often conflated with psychopaths, who "are *of essence* incompatible with rational science because they do not meet and have never met Western civilization's standards of rational personhood" (2011: 536). The language used to describe such characters in the series—including "mad men," "insane," "crazed maniac," and "murderous lunatic," for example—highlights this identification (Orthia 2011: 536). But if such language is an unambiguous signifier of the mad scientist in *Doctor Who*, what are we to do with the Eleventh Doctor's pronouncement to Amy Pond in "The Eleventh Hour" (2010), "I am definitely a madman with a box"?

The mad scientist is certainly not the only negative stereotype of scientists in popular culture. The common typecast of a scientist as a balding, middle-aged, Caucasian male wearing a white lab coat and glasses is so pervasive in society that it is even commonly reflected in the drawings of children. When asked to sketch a female scientist, children's images vary little from this basic depiction. For example, a British study found that children tend to picture female scientists as "thick glasses, flat shoes, big feet, judo types with muscular calves and sensible clothes," often referred to as the "flat chested flat heeled syndrome" (Measor and Sikes 1992: 74–75). These stereotypical images become ingrained at a young age, and while drawings of female scientists by young children have become more common in such studies in recent decades, children seem to retreat to assuming most scientists are male as they get older (Miller et al. 2018). For example, a 2006 survey of psychology majors demonstrated that college students hold similar stereotypes of scientists, drawing them as overwhelmingly male (Thomas 2006).

The insinuated unusual nature of women in science is further highlighted by the lop-sided emphasis placed on their physical appearance. In their analysis of scientists profiled in a dozen national UK newspapers from January to June 2006, Chimba and Kitzinger (2010: 612) found that nearly half of the articles made direct reference to female scientists' "clothing, physique and/or hairstyle," more than twice the frequency of their male colleagues. In the face of such microscopic scrutiny, female scientists can feel pressured to both conform to the aforementioned "flat chested flat heeled syndrome" and rein in their natural passion for their field in an attempt to prevent their appearance from detracting from their scientific credibility (Pronin et al. 2004). For example, on American network television, the antisocial and solemn Sara Sidle of *CSI: Crime Scene Investigation* (2000–15) dresses in a white lab coat over a plain white shirt and dark pants and eschews makeup.

One of the character traits analyzed by Orthia was appearance, specifically whether or not the scientists of *Doctor Who* "don the trappings of science, giving them the appearance of a scientist," most commonly signified by wearing a white lab coat and or clutching a clipboard (Orthia 2010: 231). She found that women were "moderately more likely to look like scientists than men" however in "some cases looking like a scientist is the only thing that grants these characters any scientific credibility at all, and so is tokenistic" (242). In an earlier work (Larsen 2018) I analyzed the depictions of four female scientists in the "new Who" era and noted several stereotypes. Geologist Dr. Nasreen Chaudhry's practical work boots, stereotypical black-rimmed glasses, functional black pants, button down shirt, and comfortable black women's coat paint her as an only slightly feminized version of the standard trope of scientist as nerd ("The Hungry Earth/Cold Blood," 2010). Within UNIT the brilliant but myopic asthmatic Petronella Osgood is contrasted with her superior, UNIT Head of Scientific Research

Kate Stewart. Tall, slender and beautiful, the blonde Kate dresses in tight black trousers and heels with a fitted pea coat over an equally well-fitted black woman's suit jacket and feminine camisole, while the shorter and frumpier Osgood, her brunette hair pulled back into a severe ponytail, initially wears a white lab coat, practical shoes, and ill-fitted jeans and the long Fourth Doctor scarf ("The Day of the Doctor," 2013). But no female scientist's appearance is more centrally highlighted than the fashion-conscious and sexually confident River Song. Described by the Doctor as "Hell in high heels," she is as equally comfortable in battle fatigues as she is in an evening gown ("The Wedding of River Song," 2011). Indeed, her favorite weapon to use against men (besides her sexuality itself) is hallucinogenic lipstick.

In the case of these four "noble" scientists we see two examples of the more masculine "nerd" stereotype, one fashionable but understated "power suit," and one hyper-feminized depiction. These results parallel the findings of Orthia's pre-2009 study that women in *Doctor Who* tend to conform to stereotypes of scientists' appearance, and when they don't they "break the mould [...] in strongly feminized ways" (Orthia 2010: 242). In the case of three specific recurring female scientific villains—the Rani, Madame Kovarian, and Missy—an increased emphasis on their gender and their "strongly feminized" physicality is further enhanced by aspects of the "monstrous mother" trope (Creed 1993) and "mismanaged emotions," a confluence of negative stereotypes of female scientists.

The Rani

The Time Lady known as the Rani is a highly attractive genius chemist who evokes both awe and exasperation from the Doctor and the Master on numerous occasions. For example, in the novelization of the 1987 serial "Time and the Rani," the Doctor considered the Rani "to be more brilliant than himself: a compliment he was reluctant to pay since the Rani's brilliance was devoted to the pursuit of scientific knowledge regardless of its repercussions upon man or beast, or any other species she encountered in the Universe" (Baker and Baker 1987: 8). He also sums her up as "an abomination: a brilliant but sterile mind.... There's not one spark of decency in her" (Baker and Baker 1987: 51–52).

Orthia (2010: 245) classifies this "most scientifically credible" Time Lord as "actively engaged in scientific activity on screen" and having "specialist scientific knowledge or an expert skill that is recognised by others," but she does not "look like a scientist" (232). In the Colin Baker serial "The Mark of the Rani" (1985) the inhumane researcher hides her nefarious activities within the chaos of the Luddite riots of 1820s England. She harvests the chemical responsible for sleep from the brains of local coal miners and controls selected individuals with a genetically engineered parasitic worm. She requires the chemical to counter the unintended side effects of experiments she had conducted on the citizens of the planet Miasimia Goria, her private fiefdom, a hint of the "helpless scientist trope."

In hiring Kate O'Mara for the role, John Nathan-Turner cast a well-known actress already associated with glamor and sexuality. Earlier in her career, she played in the Hammer Horror film *The Vampire Lovers* (1970) and in a highly erotic scene that focused on her cleavage was seduced by a female vampire (Cotter 2014). She appeared in numerous soap operas both before and after her time on *Doctor Who*, and her obituary in *The Guardian* called her "kittenish and seductive" (Coveney 2014). The serial's novelization highlights the Rani's physical appearance, emphasizing the "unblemished skin and sculptured beauty of a woman in her prime. Her most striking feature was her eyes; two glittering sapphires, they projected an icy calculation unflawed by compassion" (Baker and Baker 1986: 39). Valerie Frankel (2018: 104–5) terms her an "Evil Ice Queen" and "beautiful, sexy, and intelligent." She coldly kills her human assistants with as much emotion as "swotting flies" (Baker and Baker 1986: 83), leading the Doctor's companion Peri to question whether this mad scientist has any conscience. The Doctor opines, "Like many scientists […] the Rani sees us as walking heaps of chemicals. There's no place for the soul in her scheme of things" ("The Mark of the Rani"). The Rani considers her actions to be completely logical, and calls out the Doctor's inconsistent compassion in condemning her experiments while failing to defend food animals such as sheep.

Despite her immorality, she engages in a recognized scientific process toward a very specific scientific end. In "The Mark of the Rani," the Rani's misuse of science and technology is contrasted with the actions of another character, George Stephenson, a real-world pioneer of the Industrial Revolution. She is also contrasted with the Master and his stereotypical obsession with world domination; she calls him "unbalanced." Here we see the language of mental illness coupled with a lack of reason, tagging the Master as a scientific psychopath. When the Master offers her unlimited worlds to rule she notes that she already rules a planet, which is apparently more than enough for her ambitions. The Master is openly disdainful of her lack of pride when she wants to slink away from the Doctor in defeat, but she counters with "I'm a scientist. I've calculated the odds. And they and not idiotic pride dictate my actions," a notable departure from the megalomaniacal pride associated with mad scientists. However, when one of her weapons transforms a miner into a tree, she argues that he is better off in his current state as a long-lived tree. The Doctor is horrified, offering that the Time Lords should have "locked you up in a padded cell," signifying his belief that she is quite literally insane. The Rani therefore initially represents an extreme (unhealthy) case of the hyper-rational stereotype previously mentioned.

There are at least two important references to her gender within the serial. Upon retrieving her TARDIS she sheds the frumpy period clothes she had worn while disguised as an old woman and embraces her true identity as a Time Lord, revealing tight fitting leather pants and a colorful, very feminine, blouse. As she and the Master escape, one of her experiments, a T-rex embryo, begins to grow exponentially due to the Doctor's meddling with her TARDIS. The Master condemns her as a "blundering woman!" a particularly interesting statement that places the blame with her gender rather than her ineffectiveness as a scientist. This incident mirrors another "helpless scientist" incident from her youth, when her hideous genetically engineered mice ate the Lord President's cat and attacked him as well. When reminded of this by the Master, she brushes it off, citing the unavoidable potential for unexpected experimental results. Both failed experiments focus on a technological manipulation of reproduction, painting the Rani in the role of monstrous mother.

In the Sylvester McCoy era serial "Time and the Rani" her scientific ethics suffer further decline, in both enslaving the peaceful citizens of the planet Lakertya, and connecting the minds of kidnapped geniuses to a giant disembodied brain. What has not changed is her sense of fashion. The novelization describes her as a "vision in scarlet. Tight trews hugged svelte hips before tapering into knee-length boots. A shimmering brocade jacket, its stiff-edged epaulettes trimmed in gold, was belted into a slender waist before flaring into a peplum. Long brunette tresses framed a beautifully sculptured face" (Baker and Baker 1987: 7–8).

Initially she displays the same logical detachment of her previous appearance, for example equating her experimentation on "inferior species" with the Doctor's apparent lack of concern for crushing insects as he walks. The revelation that she plans to turn the planet into a monstrous "time manipulator" that will allow her to bend time to her desires signifies the completion of her metamorphosis from inhumane researcher to mad scientist, as she proudly pontificates how she will replace the chaos of the universe with order (as she defines it), including redirecting the path of evolution. Again we see the Rani labeled a monstrous mother, intent on interfering in the natural cycle of birth and death (of species in this case). Doctor finally calls her out as a "psychopath" with "murderous intent," and we see that the originally hyper-cold and logical Rani has succumbed to her passions in the end, a pendulum swing across the unhealthy emotional spectrum of extremes described by Nicholes (2012).

Madame Kovarian

Orthia (2011: 535) argues that while in the standard mad scientist trope "madness is characterized as the product of unchecked scientific obsession," within the universe of *Doctor Who* "the psychopath trope applies equally to most 'mad scientists,' rendering them mad not through scientific obsession but through mental disease." In the case of Madame Kovarian, the mental disease is religious extremism. The final episode of the Eleventh Doctor, "The Time of the Doctor" (2013), reveals that the mouthless Silents, whom no one can remember seeing once they look away, are confessional priests of the Church of the Papal Mainframe. Those whom the Doctor and his companions had battled in previous episodes were members of a splinter group called the Kovarian Chapter after its leader, Madame Kovarian, who nefariously plotted to prevent the Doctor from rescuing the Time Lords from a pocket universe by manipulating his time line.

One of Kovarian's most significant victims is Melody Pond/River Song, who is kidnapped at birth, brainwashed, and turned into an assassin who nearly kills the Doctor. Madame Kovarian is Victor Frankenstein to River Song's monster, but Kovarian's creation is ultimately redeemed. River transfers her Time Lord–like regenerations to the same man she had been designed to kill and falls in love with him, and he becomes the driving positive force in her life. For example, she achieves a doctorate in archaeology so that she can follow him through space and time. While her motivations are initially selfish, by the end of her life she has transformed from brainwashed assassin to noble scientist and gives her life to save others, including the Doctor (Larsen 2018).

The dehumanizing language used by Kovarian when she kidnaps the newly minted Doctor Song is particularly illuminating in its Frankensteinian reference to creation in a biological (natural) versus technological (monstrous) sense. She refers to the Silents as River's "owners," and taunts River, sneering "they *made* you a Doctor today, did they? [...] I *made* you what you are" ("Closing Time," 2011; emphasis mine). Kovarian clearly represents a perversion of the nurturing nature of motherhood. Previously she had ripped the newborn Melody Pond from her birth mother's arms and raised her in isolation and fear, trapping her in the technological womb of a modified space suit. Later, an adult River Song is similarly "rewombed" within a space suit she cannot control and forced by her monstrous mother to "murder" her husband.

The space suits are just one example of the technology that Kovarian integrates into her nefarious plans. Others include the Flesh (biological material used to create replicants in "The Rebel Flesh/The Almost People," 2011), time travel technology, and especially her trademark black eyepatch, an "eye drive" that allows the wearer to remember the Silents ("The Wedding of River Song"). Coupled with her tight, fitted black leather jacket and skirt, and donning severe make-up, Madame Kovarian, like the Rani, subverts the mousy "flat chest, flat shoes" stereotype previously described. In fact, she has more in common with stylish James Bond villains. Villains with eyepatches are widely seen in popular media, including Elle Driver of the 2003–04 *Kill Bill* series, Emilio Largo of 1965's *Thunderball*, and *The Walking Dead*'s Governor (the featured villain from 2012 to 2015). It should be noted that Kovarian doesn't create the technology she uses, but instead parasitically co-opts that of other species. The one, most important, exception is River Song. Kovarian's mad scientist intention is clear— she means for River Song to be used as a weapon of mass destruction against her perceived enemy. But despite her objectification in the mind of Kovarian, River Song, like the eye drive, cannot be controlled. In the end Kovarian is tortured by her own failsafe, the eye drive's debilitating and sometimes fatal electrical shock ("The Wedding of River Song"). This failure could paint Kovarian as simply a foolish or helpless scientist (in the system of Haynes), similar to nuclear physicists and their "atomic bombs" in numerous science fiction works from the 1950s such as *Them!* (1954), *Godzilla* (1954) and *Attack of the 50-Foot Woman* (1958). That is, if one could overlook her sadistic treatment of Melody Pond/River Song and her family.

Finally, in Madame Kovarian we also witness the uncontrollable emotions cited by Nicholes (2018), writ large. In fact, the stereotypical maniacal laughter of the mad scientist would have been decidedly in character for her. Madame Kovarian's "madness" differs significantly from that of the Rani, especially in its source. She leads a splinter group of the Church of the Papal Mainframe, with the schism caused by a debate over "doctrine" (how to prevent the Doctor from contacting the Time Lords). Kovarian is clearly "married" to her cause, and her rebel Silents and soldiers represent her family. The distortion of her "faith" is aligned in the parody of a family she creates. Unable to produce a child with the necessary Time Lord properties from her own flesh (a perversion of infertility), she monstrously transgresses the sacred boundaries of motherhood by ripping a newborn infant from her mother's embrace (Edge 2015). She has much in common with Dr. Mehendri Solon of "The Brain of Morbius" (1976), who tries to transplant the brain of the evil politician and cult leader Morbius into a new body, another monstrous birth. Orthia (2011: 534) explains that the zealot "traded his respectable interest in science, endorsed by the Doctor's appreciation of his work, for evil political ambition of a quasi-religious variety." Similarly, Crome (2018: 8) points out the similarity between the Kovarian Chapter and "apocalyptic terror groups."

Missy

A one-woman terror group appears in the Whoverse in the female incarnation of the Master known as Missy. In her earlier analysis, Orthia (2010: 198) terms the Master a "mad scientist" but notes that he is among the characters "represented in such a way as to shift responsibility for their actions *away from* science" (212). She argues that although his "goal was usually world/universe domination, his scientific prowess is largely incidental" (Orthia 2010: 110). I would argue that in the twenty-first century episodes the Master's technological, if not scientific, prowess is very much front and center, beginning with the Utopia Project of Professor Yana that tried to save the human species from the impending heat death of the universe ("Utopia," 2007).

The Master's backstory on Gallifrey is explored at the end of the Tennant era, where the villain's madness is revealed as the result of the childhood trauma of looking upon the Untempered Schism coupled with the Lord President Rassilon's implantation of the incessant and maddening drumming sound of a Time Lord heartbeat into his head ("The End of Time," 2009–10). At the Master's next appearance in the series in the Peter Capaldi era as the Time Lady Missy, her prim and proper Edwardian attire (complete with Mary Poppins–like umbrella) are counterposed with her over-the-top persona and at times exhaustingly emotional attachment to the Doctor. Her latest plot for universal domination takes technology to a truly mad level, uploading the minds of dying humans into a Gallifreyan matrix data-slice, tricking them into thinking that they are in an afterlife before downloading their consciousness (scrubbed of all bothersome emotions) into a Cyberman frame. This monstrous conception (taking place in the womb-like Nethersphere) gives rise to equally perverse progeny.

Like her male regenerations, Missy kills her allies and associates without blinking her now well mascara'd lashes. Most notable is her unprovoked murder of nerdy female scientist and fan favorite Petronella Osgood (or her Zygon double). After vaporizing the young woman she taunts "Thanks for being yummy" before cruelly crushing Osgood's trademark eyeglasses under her rather fashionable boot ("Death in Heaven," 2014). As a parting shot, she causes another of the Doctor's friends, UNIT chief scientist Kate Stewart, to be sucked out of the plane's cargo bay, then directs her Cybermen to destroy the plane and, randomly, Belgium. When she next appears in "The Magician's Apprentice" (2015) Missy seeks Clara's help in saving the Doctor, but when her motivations are questioned, she kills two soldiers in cold blood just to prove that she hasn't "turned good." The question is, who is she trying to prove the point to—Clara, or herself?

Missy's road to potential redemption begins in "Extremis" (2017), begging for her life when condemned by alien executioners to die for her numerous crimes. The Doctor intervenes, commuting her sentence to 1000 years in prison under his watchful eye, and afterwards acts as her therapist, friend, and mentor, while Missy appears to realize the error of her ways. Missy's gender is in the forefront of the plotline. For example, in "The Lie of the Land" (2017) Bill is amazed when she meets the mysterious and supposedly dangerous prisoner, expecting a monster rather than merely a woman. The Doctor suggests that Missy is, indeed, a monster, "going cold turkey from being bad." The insinuation that being a mad scientist is an addictive rather than simply obsessive behavior is interesting and suggests that it is a condition that can be managed but never cured.

Over the season, Missy's apparent transformation is frequently punctuated by the shedding of uncharacteristic tears. Charlie Jane Anders (2017: n.p.) argues that the audience wouldn't spend

> this much time watching Missy cry and being told to wonder if she's really changed if the Doctor's arch-nemesis were still a man. Missy's female body seems to be the main reason why this is even a dramatic point, as far as I can tell. Her tears, her insistence that her conscience is tormenting her, rely almost entirely on Michelle Gomez's use of notes of feminine vulnerability and softness.

Anders (2017) further points out that Missy is originally introduced as a "sexually aggressive older woman with a complicated past." Indeed, in "Dark Water" (2014) Missy is clearly depicted as a sexual predator, aggressively kissing the shocked Doctor when they first meet. By the end of "The Eaters of Light" (2017) Missy is not only depicted as submissive to the Doctor, but emotionally dependent upon him, craving his approval. She has, in a sense, become an eater of light herself, with an insatiable, unhealthy hunger for the Doctor's moral "light" and approval, yet another female scientist who is consumed by her emotions.

Missy's arc comes full circle in "World Enough and Time/The Doctor Falls" (2017), where we see both sides of Missy's persona, the nefarious vixen and the vulnerable subordinate. Missy is reunited with her previous (male) incarnation and manipulates him with unsettlingly incestuous flirtation before rescuing the Doctor. She is ultimately asked to choose a side, and is visibly torn, especially when the Doctor offers the absolution she has so long craved, believing she has truly changed. The Doctor never learns that in the end Missy does choose his side, but at the cost of her life—and that of her previous incarnation. Missy's transition toward being good is unfortunately seemingly concomitant with her ceasing to actually do anything scientific. Having once been a mad scientist, she can apparently become "un-mad," but can never attain the rank of a true—read *noble*—scientist. This is particularly troubling when compared with River Song's arc, in which, as previously noted, she transitions from psychopath to self-centered scientist and ultimately to noble scientist. Of course, neither River Song's nor Missy's transformation is possible without the influence of the Doctor, and again, comes at the cost of their very lives.

Conclusion: A "horrible parody of femininity"

Each of these three female scientific villains stars in their own cautionary tale against the dangers of unbridled destructive passions coupled with the technological perversion of motherhood. Their physical appearance further represents a parody of the extreme feminine. The Rani is as much mad scientist as she is fashion model, while Missy turns the prim and proper trope of the Edwardian nanny on its head. Madame Kovarian's eyepatch adds to her stereotypical representation of a woman of mystery, a look recently adopted by singer Madonna for her *Madame X* album. The unholy trinity therefore serves to illustrate Chimba and Kitzinger's (2010: 614) admonition that undue emphasis on the appearance of women scientists may result in an "implicit accusation that she is being manipulative and using her sexuality to attract attention."

Lindy Orthia reflects that, "For *Doctor Who*, sanity equates to rationality equates to science, while madness is as great a sin as superstition" (2010: 192). All three of these women scientists ultimately abandon rationality and embrace madness, albeit in different ways. In the case of Madame Kovarian, it is religious extremism and hatred for the Doctor, while for the Rani it begins as a hyper-rational single-minded focus on a misplaced goal to enhance her subjects on Miasimia Goria and ends in the stereotypical megalomania of wanting to mold the universe to her concept of what is best. Neither of these characters seems capable of redemption, as they have no concept that their actions are anything other than right.

Missy's arc is far more complex, evolving from the classic mad scientist of her male incarnations to an almost parodic, over-the-top depiction worthy of Pearl Forrester of *Mystery Science Theater 3000*. Valerie Frankel (2018: 108) sums Missy up as a "horrible parody of femininity as well as the rapacious evil matriarch and seductress in one." In "Death in Heaven" Missy floats down under her umbrella from the destruction of the UNIT plane, simultaneously offering a caricature of the beneficent Disney version of Mary Poppins and a nod to the more sinister nanny of the original books (Harmes 2015). She tempts the Doctor by presenting him with her "children" (progeny in the sense of Frankenstein or Doctor Moreau) to command, an army of Cybermen to serve the Doctor in his battle to do good. In utterly rejecting this offer, the Doctor comes to the realization that he is merely "an idiot, with a box and a screwdriver. Just passing through, helping out, learning." Note the shift from the Matt Smith era identification as a "madman with a box." The Doctor learns the offered lesson that he has been too violent and too angry, but Missy is far slower to rehabilitate, admittedly too little, too late, and only at the cost of killing her male self (the symbolism of which is certainly worthy of a separate paper).

Is Missy's rehabilitation only possible because of her female regeneration? That appears to be the message of her arc. Such a blatant gender stereotype, along with the exaggerated sexual tension exhibited towards both the Doctor and her prior incarnation, are perhaps not unexpected given the controversy over female characters in the Moffat era (Frankel 2018; Porter 2013). This debate was fueled, in part, by a rather vociferous discussion among fans and scholars as to whether or not Steven Moffat's writing (especially of female characters) reflected an inherently misogynist viewpoint (Hills 2018).

Regardless of the intention, in the end the audience is left with a number of troubling messages concerning women in science. Recall that villains as well as heroes and neutral protagonists represent the most influential characters in popular media. Indeed, in their study of scientists in *Doctor Who* through 2013 Orthia and Morgain (2016: 92) came to the unsettling conclusion that in the series, "women and men are encouraged to participate in science and compete for success, but only if they play by, accept, and commit to, masculinist rules." In clinging to the most extreme female stereotypes, the Rani, Madame Kovarian, and Missy ultimately fail as scientists.

It is instructive to remind ourselves that, for all its adult themes, *Doctor Who* was initially conceived to be a children's show. Therefore, negative stereotypes of female scientists reflected in science fiction media for children may have a chilling effect on girls who might otherwise consider science as a career, especially since research suggests the first significant drop in girls' interest in science occurs around age 10–12, the so-called tween years (Saucerman and Vaszquez 2014). While it can be argued that no mad scientist should be expected to act in anything but a cautionary role, the extreme stereotypes embodied in the Rani, Madame Kovarian, and Missy, coupled with the gender backlash surrounding the Jodie Whittaker incarnation of the Doctor (Bartlett 2019) give us pause to contemplate the deeper messages of such overtly negative depictions of women scientists, even female scientific villains.

References

American Institute of Physics (n.d.). Percent of Physics Bachelors and Ph.D.s Earned by Women, Classes of 1975 through 2016. www.aip.org/statistics/data-graphics/percent-physics-bachelors-and-phds-earned-women-classes-1975-through-2016 (accessed 18 December 2019).
Anders, C. J (2017, June 20) Doctor Who's Missy is way better when she's being bad. *Tor.com*. www.tor.com/2017/06/20/doctor-whos-missy-is-way-better-when-shes-being-bad/ (accessed 20 October 2019).
Baker, P., and Baker, J. (1986) *The Mark of the Rani*. London: W.H. Allen & Co.
Baker, P., and Baker, J. (1987) *Time and the Rani*. London: W.H. Allen & Co.
Bartlett, M. (2019) New Dimensions: Representation and allegory in Doctor Who. *Screen Education* 94: 62–69.
Chimba, M., and Kitzinger, J. (2010) Bimbo or Boffin? Women in Science: An analysis of media representations and how female scientists negotiate contradictions. *Public Understanding of Science* 19(5): 609–624.
Cotter, R.M. (2014) *The Women of Hammer Horror: A Biographical Dictionary and Filmography*. Jefferson: McFarland.
Coveney Ml (2014, March 30) Kate O'Mara obituary. *The Guardian*. www.theguardian.com/tv-and-radio/2014/mar/30/kate-omara (accessed 18 December 2019).
Creed, B. (1993) *The Monstrous-Feminine: Film, Feminism, Psychoanalysis*. New York: Routledge.
Crome, A. (2018) Plugging into the Papal Mainframe: The political role of the Church in Steven Moffat's *Doctor Who*. *Journal of Popular Television* 6: 213–226.
Edge, B.W. (2015) "She needs more": The Villainization of Infertile Women in Horror. In: Bohlmann, M.P.J., and Moreland, S. (eds.) *Monstrous Children and Childish Monsters*. Jefferson: McFarland, pp. 42–60.
Flicker, E. (2003) Between brains and breasts—women scientists in fiction films. *Public Understanding of Science* 12: 307–318.
Frankel, V.E. (2018) *Women in Doctor Who: Damsels, Feminists and Monsters*. Jefferson: McFarland.
Geena Davis Institute on Gender in Media (2018) *Portray Her: Representations of Women STEM Characters in Media*. seejane.org/wp-content/uploads/portray-her-full-report.pdf (accessed December 18 2019).
Haran, J., Chimba, M., Reid, G., and Kitzinger, J. (2008) *Screening Women in SET: How Women in Science, Engineering and Technology are represented in films and on television*. Bradford: UK Resource Center for Women in Science, Engineering and Technology (UKRC) and Cardiff University.
Harmes, M. (2015, November 26) Nasty Nannies in *Doctor Who*. *CST Online*. cstonline.net/nasty-nannies-in-doctor-who-by-marcus-harmes/ (accessed 19 December 2019).
Haynes, R. (2003) From alchemy to artificial intelligence: Stereotypes of the scientist in western literature. *Public Understanding of Science* 12: 243–53.
Hills, M. (2018) How is popular television "political"?: From the texts of Steven Moffat's *Doctor Who* to brand/fan politics. *Journal of Popular Television* 6: 167–82.
Larsen, K. (2018) The river, the rock, the relative, and the returned: Depictions of women scientists in *Doctor Who's* Moffat era. In: Carlson, A.L. (ed.) *Women in STEM on Television*. Jefferson: McFarland, pp. 187–206.
Measor, L., and Sikes, P. (1992) *Gender and Schools*. London: Cassell.
Miller, D.I., Nolla, K.M., Eagly, A.H., and Uttal, D.H. (2018) The Development of Children's Gender-Science Stereotypes: A Meta-analysis of 5 Decades of U.S. Draw-A-Scientist Studies. *Child Development* 89.6: 1943–1955.
Nevins, J. (2011, April 21) From Alexander Pope to "Splice." *Gizmodo*. io9.gizmodo.com/from-alexander-pope-to-splice-a-short-history-of-the-5794436 (accessed 20 October 2019).
Nicholls, E. (2018) A bad case of the feels: Emotion versus reason on *Blindspot*. In: Carlson, A.L. (ed.) *Women in STEM on Television*. Jefferson: McFarland, pp. 120–133.
Orthia, L.A. (2010) *Enlightenment was the choice*: Doctor Who *and the Democratisation of Science*. Ph.D. Thesis, The Australian National University.

Orthia, L.A. (2011) Antirationalist critique or fifth column of scientism? Challenges from *Doctor Who* to the mad scientist trope. *Public Understanding of Science* 20: 525–542. doi:10.1177/0963662509355899.

Orthia, L.A. (2019) How does science fiction television shape fans' relationships to science? Results from a survey of 575 *Doctor Who* viewers. *Journal of Science Communication* 18: A08. doi:10.22323/2.18040208.

Orthia, L.A., and Morgain, R. (2016) The gendered culture of scientific competence: A study of scientist characters in *Doctor Who* 1963–2013. *Sex Roles* 75: 79–94. doi:10.1007/s11199-016-0597-y.

Porter, L. (2013) Chasing Amy: The Evolution of the Doctor's Female Companion in the New *Who*. In: Leitch, G.I. (ed.) *Doctor Who in Time and Space*. Jefferson: McFarland, pp. 253–67.

Pronin, E., Steele, C.M., and Ross, L. (2004) Identify bifurcation in response to stereotype threat: Women and mathematics. *Journal of Experimental Social Psychology* 40: 152–68.

Saucerman, J., and Vasquez (2014) Psychological Barriers to STEM Participation for Women Over the Course of Development. *Adultspan Journal* 13(1): 46–64.

Steinke, J. (2005) Cultural Representations of Gender and Science: Portrayals of female scientists and engineers in popular films. *Science Communication* 27.1: 27–63.

Thomas, M.D. (2006) The Draw a Scientist test: A Different Population and a Somewhat Different Story. *College Student Journal* 40(1): 140–8.

Maxtible's Mirrors

Victorian Science in Classic-Era Doctor Who

MARCUS K. HARMES *and* RICHARD SCULLY

The 1967 *Doctor Who* story "The Evil of the Daleks" is bisected in setting and tone. Part way through its seven-part narrative, the established setting of a Victorian country house near Canterbury in the year 1866 jumps suddenly and jarringly to the futuristic interiors of the alien planet Skaro. In the narrative, a similar juxtaposition of the old and the new takes place regarding science. Deep in the Victorian manor house, a wealthy gentleman scientist is conducting experiments that are poised between the old and the new and the credible and incredible. This scientist—Theodore Maxtible—hopes to transmute base metal into gold and his alchemical practice (even for the nineteenth century) is hopelessly outmoded, having been relegated to pseudo-science since the century before (Principe 2011). But he is also attuned to the latest (for 1866) scientific thinking, including the experiments of Michael Faraday (1791–1867) and James Clerk Maxwell (1831–79) with electricity and magnetism, and Maxtible's own experiments with static electricity have proven successful.

This essay explores the resort made to the Victorian era in "Classic-era" *Doctor Who* (1963–89), and the dramatization and presentation of science in this period, which—perhaps more than any other—"permeates *Doctor Who*" (Harmes, 2014: 87). Taking as its focus science as a practice and the act of conducting science, it examines what perceptions of the nineteenth century offered to makers of twentieth century television drama and what expectations of science and scientists are captured. The primary focus is on the science in "The Evil of the Daleks," which is a heady mixture of the occult (mesmerism and alchemy), the empirical (electromagnetism and static electricity) and alien (the Daleks, time travel, and the quest for the "human factor" and the "Dalek factor").

Much of the literature on Victorian science speaks to binaries:

confident vindicated science or pseudo-science (Karpenko and Claggett 2017), the amateur and professional (Meadows 2004), the clash between religion and "the creed of science" (Lightman 2014), and the emergent mainstream compared to "popular" or "fringe" work, especially mesmerism (Winter 1994, 1998; O'Connor 2009; Hughes 2015). This essay suggests that the fictional but well-deployed conceptualization of Victorian science in "The Evil of the Daleks" speaks to a level of caution regarding these binaries. The writing, set and production design and performances in "The Evil of the Daleks" position Maxtible at the borders of practice and knowledge. He is a gentleman amateur in an age when professionalization was dawning and an earlier diversity of scientific practitioners was diminishing. He is an alchemist at a time when the borderlines between science and pseudo-science were hardening and some fields of knowledge were vindicated and others discredited. Above all else, he carries out research that is described in dialogue and presented on screen as evoking actual, empirical Victorian knowledge and scientific speculation, but which is pushed and stretched to find the sinister and alarming. In "The Evil of the Daleks," the theories and equations of James Clerk Maxwell are themselves transmuted through Maxtible's science into the truly fantastic.

Alchemy and an Apocalypse: Science in "The Evil of the Daleks"

"The Evil of the Daleks" ended *Doctor Who*'s fourth season and the seven-episode adventure (broadcast from May to July 1967) also brought the Doctor's most famous and daunting enemies back into the series. David Whitaker's script spans markedly different settings, characters and periods. The story begins in the present-day (1967) in London at Gatwick Airport and ends far from Earth on the planet Skaro. By the end of the story, the planet Skaro and the Daleks are dead, destroyed in a civil war.

However, most of the story takes place in a Victorian country house in Kent in 1866. The setting was partly recorded at the BBC's Lime Grove studios but some location filming also evoked the Victorian age using an actual Victorian structure—the country house Grim's Dyke—as the home of Mr. Theodore Maxtible, a Victorian gentleman and scientist to be sure, but also "a ruthless entrepreneur [...] obsessed with the alchemist's dream" (Howe and Walker 1997: 127). Selected for use by production assistant Timothy Combe (Guerrier 2017), Grim's Dyke is quintessentially Victorian, designed by Norman Shaw (1831–1912) and once owned by the librettist W.S. Gilbert (1836–1911). Used as a location to portray the interior and grounds of a country house in 1866, the building's dark gothic corridors, hall, minstrel gallery and carved staircase fit the period. The house strikingly juxtaposes with the alien Daleks, who glide through its corridors and menace its inhabitants; indeed, Deborah Watling—who portrayed Victoria Waterfield—recalled how her large Victorian crinoline "mirrored that of the Dalek skirts" (Watling 2010: 48). Maxtible, the master of the house, informs the Doctor that "I have the money to indulge my whims" in scientific research, and he evidently lives as a man of leisure in a large house with a staff of servants. Marius Goring, the actor playing Maxtible, is costumed and be-wigged and made up to resemble something of a cross-over between James Clerk Maxwell, Karl Marx (1818–83), and Anthony Trollope (1815–82). His somewhat demented performance echoes the scientist Rotwang in Fritz Lang's 1927 film *Metropolis* (in many ways the archetypal film "mad professor"); as does, visually, the scenic design of the scientific apparatus set against gothic portals, managed by series designer Chris Thompson (Howe and Walker 1997).

Inside Maxtible's laboratory, the production design created the same contrast of the Victorian with the alien, enhancing the BBC's reputation for excellence in period drama, and "evoking impressions of sources as diverse as H.G. Wells and Lewis Carroll" (Howe and Walker 2003). Although the original realization of the scene from Episode 2 (the only one of the seven original episodes to have survived) is rather less spectacular, John Peel described the setting in his 1993 novelization of the story (84):

> It was apparent that the room had once been a conservatory of some kind, and had been converted into a scientific laboratory. The walls were all wood-panelled. Previous owners of this ancient house would no doubt have cried out in horror at the damage done to the panelling with the wiring and apparatus now filling the room. At first glance, the place looked as if some glassmaker had collided with a wiring contractor. Tubing filled the room, creating vast glass archways carrying thick coils of conductive wiring. Shafts and tunnels of glass led from the many instrument-filled benches. In one corner was a primitive dynamo, though the Doctor noted that there were several von Siemens coils that could only have been invented a few months earlier.

On screen, the same effect is achieved largely through foregrounding a long, L-shaped table filled with bubbling and steaming beakers and pipettes. Maxtible's cabinet of mirrors—which connects Victorian Kent to the planet Skaro via time travel—is a carved gothic structure on the outside with shimmering technology inside:

> Most of the cables and glass tubes led to this cabinet. Though it was primarily made from wood, it had a large glass dome and metal panels. The door was lined with insulating substances. Barely visible from this angle and protruding into the glass dome were large shafts that resembled lightning conductors [Peel 1993: 84].

From time to time throughout the story, a Dalek is framed in its gothic portal, emphasizing the juxtaposition of Victorian gothic with alien design.

The science Maxtible pursues in content and method is both of the past and the future. A modern-day Faustus, his research and capacities traverse alchemy, mesmerism, and electromagnetic theory. His apparatus of mirrors and their magnetized surfaces recalls the trappings used by Franz Mesmer (1734–1815) in his pseudo-scientific sessions (Basham 1992), and the whole concept is largely an extension of C.H. Hinton's (1896) and H.G. Wells' (1897) interest in invisible planes (including the "fourth dimension") and the bending of light (Williams 2007). At his leisure, Maxtible has had time and money to indulge his scientific whims and one of these, an enduring fascination, is "the concept of traveling through time." To this end, Maxtible has theorized that "a mirror reflects an image, does it not? [...] So, you may be standing there, and yet appear to be standing fifty feet away." Pursuing that suggestion to a (for Maxtible) logical conclusion, Maxtible has placed one hundred and forty four mirrors made out of polished metal into the gothic cabinet and he further theorized, "Like repels like in electricity, Doctor [...] I attempted to repel the image in the mirrors, wherever we directed." Having used negative and positive electricity, Maxtible and his colleague Waterfield used static electricity. Their experiments have a patina that is scientifically correct and evocative of the Victorian age. Maxtible is aware of current research, and claims that he has been "following the new investigations twelve years ago by J. Clerk Maxwell into electromagnetism and the experiments by Faraday into static electricity." This is despite the fact that the real Maxwell did not publish his *Treatise on Electricity and Magnetism* until 1873, some seven years after the 1866 setting of "Evil of the Daleks" (Guerrier 2017).

But Maxtible's principal focus lies elsewhere and with a branch of learning that even by 1866 was outmoded. Maxtible informs his daughter that his research focuses on "the greatest secret of all. Some say it was known to the ancient alchemists, some say that the secret never existed at all, but still the stories and the rumours, and the search goes on." In short, he directs his wealth and energy to "The transmutation of metal into gold [...] To possess such a secret would mean power and influence beyond all imagination. And I am about to discover this secret."

The delineation of Maxtible in these terms, as wealthy, living in and owning a large country house, and conducting scientific research at his leisure, tallies with other nineteenth century scientists, professional and amateur (Guerrier 2017). William Thompson, first Baron Kelvin (1824–1907), built the baronial castle Netherhall in Ayrshire, which he powered with electricity and where he worked on experiments (Thompson 2004). For a time, James Clerk Maxwell lived and worked on his research at his own private estate. Robert Gascoyne-Cecil, third Marquess of Salisbury (1830–1903; three times prime minister, 1885–86, 1886–92, 1895–1902) who together with Lord Kelvin attended meetings of the Institute of Electrical Engineers, electrified the Tudor Hatfield House. Like Maxtible, he also had a laboratory in his home where he relished spending much of his spare time (Gooday 2015; Roberts, 1999). These houses are an important intersection of the historical with the technological. Lord Kelvin introduced electric light to his house and like other landed gentry who were also actively engaged with science and technology, his house was simultaneously an example of Victorian country house architecture that evoked instant medievalism and baronial splendor and the latest modern technology. Similarly, Cragside, the country house of artillery magnate William, first Baron Armstrong (1810–1900), was (like Grim's Dyke) a historicist jumble of chimneys, half timbering and gables. However, it was also in the vanguard of engineering and powered by advanced technology including hydroelectricity, telephony and incandescent lamps (Gooday 2015). Elsewhere on screen in contemporary shows including *The Avengers* (1961–69) and *The Saint* (1962–69) (both set in the present day), the world of the gentleman amateur was shown as reaching the end of its heyday, whereas period-set *Doctor Who* provided a framework of the gentleman amateur. Maxtible was poised between past and present, but part of narratives supposing that scientific success seemed more of an accident than an obvious outcome of the scientific milieu brought onto the screen.

The juxtaposition of the Victorian and the modern is striking but also uncanny, making the serial one of the all-time greats in the opinion of Steven Moffat (2012). Dialogue in "The Evil of the Daleks" imbues the technology—both human and Dalek—with sinister resonances. The Daleks themselves are "creations of the devil." These uncanny resonances between the static electricity and the supernatural are a further point of connection between Maxtible and actual nineteenth-century gentlemen scientists. Graeme Gooday's (2015: 121–52) study of the cultural history of electricity points out that Lord Salisbury spoke of electricity in ways that made it seem an "anthropomorphized power"; that Salisbury talked of it in prophetic terms; and he spoke of its "distributive capacity," making it a force by which "every human relation will be powerfully affected." Salisbury's sense of the disruptive as much as distributive capacities of electricity attracted criticism at the time, as social reformers such as W.T. Stead (1849–1912) found Salisbury's paeans did not extend to making electricity available for practical things like tramcars and street lighting. Salisbury's avoidance of the prosaic and his elevation of the electric to a prophetic, anthropomorphic entity is still not the intrusion of the alien into the Victorian seen in "The Evil of the Daleks." It is however a historical glimpse of a gentleman scientist amateur alert to the mysterious and even supernatural dimensions of his research.

Science and the Nineteenth Century

Maxtible's laboratory, a gothic chamber, connects across time and space to an alien world, Skaro, and the wood paneling and gothic portals of the past links with the gleaming metal of the alien world. Twentieth-century science fiction, especially in Britain, has turned repeatedly to the nineteenth century as a setting and a source for inspiration. However, the nineteenth century generated its own science fiction, most notably Mary Shelley's *Frankenstein; or, the Modern Prometheus* (1818), Robert Louis Stevenson's *The Strange Case of Dr Jekyll and Mr Hyde* (1886), and H.G. Wells' *The Invisible Man* (1897). There is obvious slippage in terminology and genre including the notions of the "scientific romance" (of Hinton 1896), and speculative fiction (see: Clarke 1992) rather than works of science fiction, although science fiction writer Brian Aldiss (1975) charts a meaningful genre trajectory from the gothic literature of the late-eighteenth century to a recognizable body of nineteenth century science fiction texts. Aldiss describes the gothic works as comprising "a dream world" from which sprang science fiction. Laura Hilton (2011) notes but expands on Aldiss's suggestion, in particular pinpointing the important distinction that the supernatural of the gothic gives way to the empirical research privileged in science fiction. Victor Frankenstein reflects that intellectual pathway, characterized as moving away from the Renaissance natural philosophy, repudiating an antique and intellectually stagnant curriculum in favor of modern branches of knowledge.

The nineteenth century generated its own science fiction but continued to inspire works in the twentieth century. The development that Hilton notes, the transition from the supernatural to the experimental, is important to understanding the value of the nineteenth century to the science fiction of the twentieth. The notion and identity of scientist, scientific apparatus, and electricity were all features of nineteenth-century scientific research that recurred in twentieth century science fiction. Notably, the trajectory from gothic to science fiction noted by Aldiss and discussed by Hilton is complicated by the twentieth century treatment of the nineteenth. Particularly in the popular Frankenstein horror films of Hammer Film Productions from 1957 to 1974, the science fiction found creative stimulus in the gothic. Surgery and electricity, poised at the vanguard of nineteenth century scientific developments, were in a *mise en scene* of gothic catacombs and castles. Hammer's iterations of the Frankenstein character and stories relocated the science fiction among these gothic trappings, in contrast to the modernist and expressionist designs in Universal's versions from the 1930s.

One further point of importance emerges from this brief consideration of Hammer, which is the counterpoise between the orthodox and the heterodox. Curiously, in Mary Shelley's original story, Frankenstein abandons what would now be regarded as the limited and misguided tenets of natural philosophy bequeathed by intellectual figures such as Cornelius Agrippa (Harmes 2015: 57). His pursuit of a more modern curriculum places him in a scientific vanguard but what he creates and his methods in pursuing his studies push him beyond the fringes of scientific respectability. The films made by Hammer, loosely adapted from aspects of Shelley's novel, toyed with and deployed these binaries of the orthodox and heterodox and the respectable and the outrageous. In their first, *The Curse of Frankenstein* (1957), Baron von Frankenstein (elevated to the aristocracy in the Hammer version, with its still-sinister wartime Germanic connotations) is socially exalted and intellectually respectable, at least at a surface level, but he suffers a fall from grace and respectability. In the sequel, *The Revenge of Frankenstein* (1958), a medical council upholds standards, and condemns and prohibits certain types of research, as well as driving out Frankenstein from their professional community.

In these films, the role of Frankenstein was played by Peter Cushing (the star of the 1965 and 1966 films *Doctor Who and the Daleks* and *Daleks: Invasion Earth 2150 AD*), and in 1959 he played a similar part, Dr. Knox, in *The Flesh and the Fiends*. In this instance, the part is based on an actual person, the Edinburgh anatomist and surgeon Dr. Robert Knox, notoriously associated with the body snatchers Burke and Hare. The film delineates the same vision of a nineteenth-century scientific researcher poised between public acclaim and the possibility of scandal and ridicule. One of the dramatic licenses taken in the film is its ending. In actuality, Knox left Edinburgh in disgrace and his medical career did not recover from the disgrace of his association with body snatching. However at the end of the film Knox returns in triumph to his lecture theatre and received an ovation from his medical students, in tribute to his courageous if amoral pursuit of his research.

This filmic context creates a meaningful background for Theodore Maxtible. These popular films reinforced a particular vision of science and scientists in the nineteenth century and showed a complex interplay of different factors. These include the rich amateur contrasted with the trained professional, the pioneer whose research imperiled his social standing compared to the stodgy respectability of a medical council or board, and the orthodox or unorthodox knowledge and methods of science (with plenty of Bunsen burners, bubbling beakers, Erlenmeyer flasks, and the like). Other horror efforts including *Blood of the Vampire* (1958) and *The Asphyx* (1972) reiterated this impression of a combination of wealth and secrecy as the basis of a gentleman scientist's work.

Maxtible, Maxwell and the Mirrors

Maxtible's research ends in disaster. His one hundred and forty four mirrors, each charged with static electricity, open a pathway for the Daleks to come to Kent from Skaro, following which they invade his house, take one of the occupants prisoner, force others to undergo the terrifying experience of time travel, and eventually blow up the house. Maxtible himself dies in an inferno on Skaro, having lost his mind and humanity when possessed by the "Dalek factor." These researches, however outlandish and drastic their outcomes, had a patina of plausible science at their foundation, largely sourced from *Encyclopaedia Britannica* and other reputable venues (Guerrier 2017). Maxtible's references to Faraday and Maxwell make the science fiction intersect with actual Victorian science. The Doctor meets Maxtible in 1866, a propitious time for someone to be interested in electromagnetism. David Whitaker's script set the Victorian sequences of the serial one century before action in the present day, but the mid–1860s were a point of maturity in electrical research. In 1860 Maxwell held King's College London's Chair of Natural Philosophy. In his inaugural lecture he noted that the "present generation has no right to complain of the great discoveries already made, as if they left no room for further enterprise" (Maxwell 2009: 667). By that, he acknowledged the field of electrical and magnetic sciences was maturing and an "immense mass of facts has been collected, and these have been reduced to order and expressed as the results of a number of experimental laws."

However Maxwell insisted that there remained a frontier of research in the field, and his comments in the actual 1860s provide important context for the fictional 1866 in "The Evil of the Daleks." Both Faraday and Maxwell theorized and imagined as well as conducting practical experiments, in much the same way that Maxtible describes his flights of speculative fancy into time travel as well as his practical tests. They also found that their practical experiments gave rise to further theoretical problems (Ronan 1983). Maxtible's interest in how his polished and charged metal plates would allow him "to refine the image in the mirror, and then to project it" was a theoretical speculation with a grain of reality in it. Faraday's experiments, referred to by Maxwell, had raised the question of "how electricity and magnetism could affect each other across empty space," as noted by the historian of science Colin Ronan (1983). As noted earlier, Maxwell's *Treatise on Electricity and Magnetism* did not appear in print until 1873, but by the mid–1860s Maxwell was engaged on developing the mathematical equations and the research that allowed him in 1861 to propose the aether as the basis of interaction between electric currents and magnetic fields (Ronan 1983).

Maxtible's science of mirrors and static electricity therefore intersect with actual Victorian science, and the story implied that Maxtible had been present at the Royal Society meeting for Maxwell's 1864 paper "A Dynamical Theory of the Electromagnetic Field" (Guerrier 2017). What therefore of Maxtible's alchemy and another branch of his practice: mesmerism? His quest to transmute metal to gold was mentioned above. Elsewhere in the story (in Episode 5) it is seen that he is an adept hypnotist and uses a faceted crystal to hypnotize his maidservant and Victoria Waterfield. His scientific knowledge gained from Faraday and Maxwell was *avant garde* for his period and has since been endorsed as credible and respectable by later generations, whereas alchemy has not and mesmerism also sits at the fringes of respectability. Mesmer himself had earlier sought the recognition and regard of the medical profession and failed to gain it (Basham 1992).

Maxtible makes no claims to professionalism; indeed as a wealthy gentleman he may well have disdained such notions. As Hannah Gay points out, the nineteenth century was still an era where the gentleman scientist working with the help of one or more assistants was still a recognizable scientific role (Gay 2008). Scientific circles in the 1850s still spoke favorably of the private and voluntary as being the natural outlet and support for the "national genius" (in Turner 1980: 591). Yet rhetoric such as that did not obscure increasing desire for and campaigning for the national endowment for scientific research, rather than relying on private associations or a gentleman's private wealth.

It was however also a role that was passing from the commonplace and familiar, giving way to a more professional body of researchers who staked a claim to professional knowledge and authority (Gieryn 1983), who operated within professional research laboratories and societies (Barton 2003), and who were expected to uphold professional standards, including the accuracy of their scientific observations (Cawood 1979). Maxtible's characterization speaks to these developments across different levels. On one level, his status and his performance of science as a gentleman amateur was a type passing from the norm. However, a body of professional scientists contributed to the dispersal of knowledge within wider society. James Clerk Maxwell was himself concerned "about the paucity of research" that was circulating, but the growing number of professional scientists meant that ideas and theories could be dispersed to an educated but popular level within British society (Turner 1978: 359). Maxtible's actions, his performance of science and his knowledge therefore evoke a complex impression of science where professionalism did not entirely eclipse the amateur but interacted with it.

Professional identities have an impact on what constituted a mainstream body of knowledge, which excluded "sea serpent investigations, phrenology, and spiritualism" (Lyons 2009: 171). Other fields of knowledge were vindicated. That impression was not of course clear-cut or absolute. For example, the participation of Darwinian science in intellectual life was defined by its complexity. As Walter Cannon observes, the aftermath of the publication of *Origin of Species* was the creation of a "multi-normative world," pitting science against religion and even creating conflicts within scientific circles, whereby "biology was pitted against physics, geology against astronomy, intuition against statistics" (Cannon 1964: 489). Cannon charts a major epistemological challenge, in which a world which had seen the alliance between religion and science gave way to a later Victorian world which "had to live with many" truths (489). The scientific world, of which Maxtible's research is a fictional distillation, was also one influenced by continental scientific practice. In doing so, the notions of boundaries and binaries are again apparent. The affront to the alliance between religion and science, which Cannon observes in relation to Darwinism, is noted in the awareness by British scientists of the dispassionate, rationalist proceeding of continental practitioners.

Awareness and application of this approach hardened some boundaries between "scientific and religious knowing, between rational and affective knowledge" (Richards 1997: 52). But the identification of binaries in Victorian science should be undertaken with caution, and Maxtible's science speaks into this caution. One consistency in Maxtible's scientific practice is that his alchemical research is practiced in private. Chrysopoeia, or alchemy, disappeared from respectable scientific circles during the eighteenth century, defined by a binary as what chemistry was not and leaving alchemy as an intellectual taboo (Principe 2011). The consequence was that those who continued to practice alchemy did so privately. Maxtible's mesmeric abilities align with the alchemy, as a partial revival of alchemy in Victorian England was coterminous with the rise of occult secret societies. David Whitaker's scripts build these associations, not only in Maxtible's talk of alchemy and his practice of mesmerism but the associations built up in dialogue of the sinister and the uncanny that surrounds him. The intrusion of the alien Daleks in the house is misunderstood by the servants as ghosts and hauntings and the location filming in and around Grim's Dyke evokes the trappings of the old dark house (Guerrier 2017). Lawrence Principe's research into the seventeenth, eighteenth and nineteenth centuries of alchemical practice again caution against perceptions of binaries. As he notes, esteemed scientific figures including Robert Boyle and Sir Isaac Newton were alchemists, aspects of their careers that the editors and custodians of their scientific papers often kept hidden (Principe 2011). In that regard, the combinations of the writings of Faraday and Maxwell, the hypnotism, the occult resonances and the alchemy that all combine in the scientific career of Theodore Maxtible take on the character of a more measured and reasonable vision of scientific history, where binaries come back together.

Coda: The Victorians and Their Science Elsewhere in Doctor Who

Maxtible was the first but not the last *Doctor Who* character in which the amateur, the wealthy, and the dangerous, cohered in a historical setting, normally Victorian. The setting of "Pyramids of Mars" (1975) is a rich scholar's large private estate. The house belonging to Professor Marcus Scarman, an archaeologist, becomes the site of intrusion of alien technology from Mars. The Edwardian (close successor to the Victorian) era's own science though is also represented, as one character—Scarman's brother Laurence—has the intelligence and leisure time combined to indulge his scientific fancies and has created a radio telescope, or as he calls it a Marconiscope (after Guglielmo Marconi), to listen for messages from outer space. He also refers to "that novelist chap, Mr. Wells" in one of the series' most common points of reference and a key nexus for the nineteenth and twentieth centuries. For a time, the Edwardian science interacts with the alien, and the Marconiscope is able to cut off the psionic control an alien on Mars is using to control the archaeologist. So too, "The Talons of Weng-Chiang" (1977)—set shortly after the Jack the Ripper killings—intruded technology into the world of Victorian music hall. On this occasion the technology, while still of human origin, was alien to the Victorian period, being instead a product of the fifty-first century brought back in time. It was, however, designed or disguised to be like Victorian chinoiserie artifacts.

The Victorian gentleman Josiah Samuel Smith in 1989's "Ghost Light"—like Maxtible—had the wealth, leisure and exploitable human resources in his household servants to conduct his own research. His home is advanced for its time with electric lighting, telephony, and a mechanical lift, in emulation of the country seats of figures such as Lords Kelvin and Salisbury. Where David Whitaker's scripts alluded to Maxwell and Faraday to place actual Victorian science into the mouth of his character, Marc Platt's script again evokes Victorian science, this time Charles Darwin. The combat between the Church of England and Darwinian scientists including Thomas Huxley plays out in "Ghost Light" with a distinctive Victorian scientific twist, when the Reverend Matthews, an Anglican clergyman who disputes the theory of natural selection, chemically and surgically devolves into an ape (Harmes 2013). Matthews is killed and mounted as a pathological specimen in a case. In Smith's Victorian villa, Gabriel Chase, the Victorian science intersects with the alien (and Smith himself is revealed as an alien creature hiding beneath a veneer of gentlemanly demeanor). Smith's laboratory contains displays of macabre pathological specimens, and the display of taxidermy, the sturdy wooden display cases and the concern with taxonomies evokes what was the burgeoning world of Victorian natural history museology and the display of specimens as scientific symbols (Yanni 2005). The plot and thematic influences of "Ghost Light" combine Darwinism, racism, ghost stories, religion and imperialism. An explorer, Redvers Fenn Cooper, is held captive in Gabriel Chase and Josiah Smith plans to use him assassinate Queen Victoria and take over both the crown and the Empire. Historians of Victorian museums note the connection between the expansion of the British Empire and the development of natural history museums, which displayed exotic items from colonized lands and peoples (Black 2000). Victorian museums were and remain colonizing spaces, in which British explorers and imperialists displayed their trophies (see Mills, this volume).

Finally, the chemist and engineer Mrs. Gillyflower in "The Crimson Horror" (2013) is the latest iteration of a rich Victorian scientist in *Doctor Who*. Although this essay focuses on the "Classic" era, her character merits attention as a reiteration of Victorian science in the series with the addition of gendered themes. Mark Gatiss's script engages with multiple layers of storytelling. Mrs. Gillyflower's horrific Victorian brutality and austerity is reminiscent of the tyrannical Mrs. Clennan in Charles Dickens' *Little Dorrit* (1857). In addition, Victorian science and industry together with the vigor of northern nonconformist religion and lurid penny dreadfuls combine as influences. Mrs. Gillyflower's ideal community Sweetville emulates the Cadburys' Bournville. Where the Cadburys' ideal community was an act of Quaker beneficence, Sweetville is wholly sinister. The community is where a few selected individuals, based on principles of survival of the fit and attractive, will be sheltered when a rocket rains down destructive poison on the rest of the world. The people she has placed in suspended animation beneath giant glass domes resemble (as did the Reverend Matthews in "Ghost Light") full-scale pathological specimens in a Victorian museum or the ornamental stuffed birds found in Victorian parlors. Viewers learn that she used her daughter as a specimen for experimentation, blinding and disfiguring her in the process. On screen, viewers learn very little of Mrs. Gillyflower's background. Socially, her accent and poor table manners betray working class origins; viewers learn little else about where she obtained the technical skills that have made her a "prize winning chemist and mechanical engineer," although the symbiotic influence of the alien creature "Mr. Sweet" may be part of her achievement.

However contextually, she is more of an amateur than even Maxtible or Smith. She is a woman in a period when women could not attend an English university (let alone take a degree). Some social reformers including Emily Davies attempted to promote scientific education for girls and women. The Schools' Inquiry Committee, the University of London examination for women and campaigns for higher education included discussion of the merits of girls and women learning science without limitations (Kamm 2012). Many of these developments post-date the fictional context of Mrs. Gillyflower. Her scientific expertise will have been autodidactic. It was possible for Victorian women to acquire scientific knowledge in the nineteenth century: cheap and accessible books for general readers appeared routinely in print (Lightman 2000). The place for women in Victorian science is complex in its character and range. Publications existed purely for female readers and ensured women could gain access to scientific knowledge, doing so with care and diversity. Available scientific literature included periodicals, textbooks and coffee table books, catering to "women of different ages, classes, and levels of expertise" (Shtier 1999: 236). Yet as knowledge increased, opportunity diminished. The same move towards professionalism that eclipsed the gentleman amateur like Maxtible also pushed women to the periphery of scientific research and activity, losing even the more marginal roles of helpmates and investigators (Shtier 1999).

Conclusion

Contextually, Maxtible was the first but not the last Victorian-era scientist in *Doctor Who*, and in some later serials discourses of discovery, conflict, experimentation and exploitation have been brought onto screen through serials featuring a nineteenth or early–twentieth century scientist. Characterization, performances, sets and themes together combine to create a memorably mad evocation of science that is embedded in striking aspects of the Victorian age and its scientific knowledge.

The interplay between success and failure speaks to the contrasting themes raised by Maxtible and his mirrors. A gentleman amateur in an age of increasing professionalization and an alchemist in an age which largely scorned the idea, Maxtible intersects with some aspects of Victorian science that have not been lauded by later generations. But simultaneously, Maxtible was abreast of recent electromagnetic theory, and the portrayal of science in this serial speaks to awareness that many divisions or binaries proposed about the era and its knowledge are not so clear cut as sometimes suggested.

REFERENCES

Aldiss, B. (1975) *Billion Year Spree: The History of Science Fiction.* Corgi Books.
Barton, R. (2003) "Men of Science": Language, Identity and Professionalization in the Mid-Victorian Scientific Community. *History of Science* 41: 73–110.
Basham, D. (1992) *The Trial of Women: Feminism and the Occult Sciences in Victorian Literature and Society.* New York: New York University Press.
Black, B.J. (2000) *On Exhibit: Victorians and Their Museums.* University of Virginia Press.
Cannon, W.F. (1964) The Normative Role of Science in Early Victorian Thought. *Journal of the History of Ideas* 25(4): 487–502.
Cawood, J. (1979) The Magnetic Crusade: Science and Politics in Early Victorian Britain. *Isis*, 70(4): 492–518.
Clarke, I.F. (1992) *Voices Prophesying War: Future Wars, 1763–3749,* 2nd Edition. New York: Oxford University Press.
Gay, H. (2008) Technical Assistance in the World of London Science, 1850-1900. *Notes and Records of the Royal Society*, 62: 51–75.
Gieryn, T.F. (1983) Boundary-Work and the Demarcation of Science from Non-Science: Strains and Interests in Professional Ideologies of Scientists. *American Sociological Review* 48(6): 781–795.
Gooday, G. (2015) *Domesticating Electricity: Technology, Uncertainty and Gender, 1880-1914.* Routledge.
Guerrier, S. (2017) *The Black Archive #11: The Evil of the Daleks.* London: Obverse Books.
Harmes, M. (2013) *Religion, Racism and the Church of England in Doctor Who.* In: Orthia L (ed) *Doctor Who and Race.* Bristol: Intellect Books, pp. 197–212.
Harmes, M. (2014) *Doctor Who and the Art of Adaptation.* Rowman and Littlefield.
Harmes, M. (2015). *The Curse of Frankenstein.* Columbia University Press.
Hilton, L. (2011) Gothic science fiction in the steampunk graphic novel *The League of Extraordinary Gentlemen.* In: Wasson, S., and Alder, E. (eds.) *Gothic Science Fiction 1980-2010.* Liverpool: Liverpool University Press, pp. 189–208.
Hinton, C.H. (1896) *Scientific Romances.* London: Swan, Sonnenschein & Co.
Howe, D.J., and Walker, S.J. (2003). The Evil of the Daleks. *Doctor Who: The Television Companion.* www.bbc.co.uk/doctorwho/classic/episodeguide/evildaleks/detail.shtml (accessed 14 November 2019).
Howe, D.J., Stammers, M., and Walker, S.J. (1997) *Doctor Who—The Handbook: The Second Doctor.* London: Doctor Who Books.
Hughes, W. (2015) *That Devil's Trick: Hypnotism and the Victorian Popular Imagination.* Manchester: Manchester University Press.
Kamm, J. (2012) *How Different from Us: A Biography of Miss Buss and Miss Beale.* Routledge.
Karpenko, L., and Claggett, S. (eds.) (2017) *Strange Science: Investigating the Limits of Knowledge in the Victorian Age.* Ann Arbor: University of Michigan Press.
Lightman, B. (2000) Marketing Knowledge for the General Reader: Victorian Popularizers of Science. *Endeavour* 24(3): 100–106.
Lightman, B. (2014) The Creed of Science and its Critics. In: Hewitt, M. (ed.) *The Victorian World.* London & New York: Routledge, pp. 449–465.
Lyons, S.L. (2009) *Species, Serpents, Spirits, and Skulls: Science at the Margins in the Victorian Age.* SUNY Press.
Maxwell, J.C. (2009) *The Scientific Letters and Papers of James Clerk Maxwell Vol. 1.* Cambridge: Cambridge University Press.
Meadows, J. (2004) *The Victorian Scientist: The Growth of a Profession.* London: The British Library.
Moffat, S. (2012) Steven Moffat on The Evil of the Daleks. *Doctor Who.* www.bbc.co.uk/programmes/p01240sf (accessed 14 November 2019).
O'Connor, R. (2009) Reflections on Popular Science in Britain: Genres, Categories, and Historians. *Isis* 100(2): 333–345.
Peel, J. (1993) *The Evil of the Daleks.* Virgin Books.
Principe, L.M. (2011) Alchemy Restored. *Isis* 102: 305–312.

Richards, J.L. (1997) The Probable and the Possible in Early Victorian England. In: Lightman, B. (ed.) *Victorian Science in Context.* Chicago: Chicago University Press, pp. 51–71.

Roberts, A. (1999) *Salisbury: Victorian Titan.* London: Weidenfeld & Nicholson.

Ronan, C.A. (1983) *The Cambridge Illustrated History of the World's Science.* Cambridge: Cambridge University Press.

Shtier, A.B. (1999) *Cultivating Women, Cultivating Science: Flora's Daughters and Botany in England, 1760–1860.* Johns Hopkins University Press.

Thompson, S.P. (2004) *The Life of Lord Kelvin, Volume 2.* American Mathematical Society.

Turner, F.M. (1978) The Victorian Conflict between Science and Religion: A Professional Dimension. *Isis* 69(3): 356–376.

Turner, F.M. (1980) Public Science in Britain, 1880–1919. *Isis* 71(4): 589–608.

Watling, D. (2010) *Daddy's Girl: The Autobiography of Deborah Watling.* London: Fantom.

Wells, H.G. (1897) *The Invisible Man: A Grotesque Romance.* New York & London: Harper & Brothers.

Williams, K. (2007) *H. G. Wells, Modernity, and the Movies.* Liverpool: Liverpool University Press.

Winter, A. (1994) Mesmerism and Popular Culture in Early Victorian England. *History of Science* 32(3): 317–343.

Winter, A. (1998) *Mesmerized: Powers of Mind in Victorian Britain.* Chicago: University of Chicago Press. Yanni, C. (2005) *Nature's Museums.* Princeton Architectural Press.

The Victorians Sleeping in Our Minds
Victorian Scientific Enquiry in Old and New Series Doctor Who

Catriona Mills

> VICTORIA: You probably can't remember your family.
> THE DOCTOR: Oh yes, I can when I want to. And that's the point, really. I have to really want to, to bring them back in front of my eyes. The rest of the time they ... they sleep in my mind and I forget.
> —"The Tomb of the Cybermen" (1967)

No period of time has fascinated *Doctor Who* like the nineteenth century (Mills 2013). This essay argues that an underexplored aspect of that fascination is the program's lingering debt to modes of scientific enquiry whose modern forms reached peaks of development in the Victorian era. In its original incarnation (1963–89), *Doctor Who* occupied a place where the familiar scientific romances of the late Victorian period were giving way to a new model of science fiction; like the First Doctor's late Victorian-Edwardian costume, the traditions of the scientific romance carry through. What is perhaps less recognized is the extent to which the new series (2005–present) continues to query Victorian scientific concerns. This essay considers three stories that deal with three Victorian scientific obsessions: dimensionality, archaeology, and museum building.

Firstly, I consider dimensionality in "Flatline" (2014) as a legacy of late Victorian scientific and science-fictional enquiry, evident in works by authors such as H.G. Wells and Edward Abbott Abbott. In examining the rise of theories around the fourth dimension in the Victorian period, I consider the ways in which *Doctor Who*, ostensibly a program to which dimensionality is central and indeed discussed in its first episode in 1963, covers new ground for the program in "Flatline." Then, I examine how the

Victorian obsession with archaeology informs "The Tomb of the Cybermen" (1967), with a focus on race- and empire-infused archaeological-adventure stories, such as the works of H. Rider Haggard. Finally, I analyze the natural partner of scientific enquiry, the Victorian mania for museum building, in "The Space Museum" (1965). The order in which the stories are analyzed accords with the directness with which they address Victorian modes of science rather than production history: in "Flatline," for example, the Victorian artifacts are actively threaded through the story by the script-writer; in "Tomb of the Cybermen," they are set at a remove by the futuristic setting but heavily influence the story-telling decisions; and in "The Space Museum," as befits museum exhibits, they may be truly said to be sleeping. Examining these stories helps to uncover and therefore question some of the ways in which Victorian scientific enquiry continues to sleep in the minds of *Doctor Who* writers and producers, to be brought back in front of their eyes as needed.

The Dimensions of Biology

Towards the end of the nineteenth century, a new scientific concern began to emerge: the fourth dimension. As Ian Stewart (2002: xix) notes, "The fourth dimension was very much 'in the air' in the late 1800s. The interest began among scientists and mathematicians, but their excitement transmitted itself to the general public." That is, what began as mathematical theories became fiction. The idea of the fourth dimension, as Deanna K. Kreisel sets out in detail (2014), arose from contemporary innovations in analytical geometry, which led to what is variously called hyperspace philosophy, n-dimensional geometries, or higher-dimensional geometries. This theorem posited the extension of a cube into the fourth dimension, to form a hypercube or tesseract (just as the extension of a square into the third dimension forms a cube). One of the primary figures in the development of ideas around the fourth dimension was Charles Howard Hinton, mathematician, coiner of the word "tesseract," and author of scientific romances. Hinton's (1897: 25) description of the potential denizens of the fourth dimension is distinctly reminiscent of the way the Doctor is perceived by secondary characters and by the audience:

> A being existing in four dimensions must then be thought to be as completely bounded in all four directions as we are in three. [...] There would be no barrier, no confinement of our devising that would not be perfectly open to him. He would come and go at pleasure; he would be able to perform feats of the most surprising kind.

The plot potentiality of what could otherwise be a relatively dry theorem appealed to the writers of scientific romances. Of these, perhaps the best known to modern readers is *Flatland* (1884) by schoolmaster and theologian Edwin Abbott Abbott: in this novel, a Square from a two-dimensional world encounters a Sphere from a three-dimensional world, who opens his eyes to not only multiple dimensions but also the social mores of his own place, in what Elizabeth Throesch (2009: 37) calls the "pinnacle [of the fourth dimension] as a fictional device." Charles Hinton wrote a series of short stories, including "A Plane World," which were published along with essays on the mathematics behind the fourth dimension in a collection called *Scientific Romances* (1884). George Macdonald posited no fewer than seven dimensions in *Lilith* (1895), in which a librarian follows his ghostly predecessor into a parallel, symbolist realm.

But the fourth dimension is perhaps less familiar to modern readers than it would have been to readers of the mid to late Victorian period. Some works that grapple with Victorian mathematics do remain popular, such as *Alice's Adventures in Wonderland* (1865) and *Through the Looking Glass* (1871), both written by a mathematics don, and which Throesch (2009: 38) argues highlight "some of the pitfalls of new developments in symbolical algebra, as well as non–Euclidean and n dimensional geometries of the later nineteenth century." Yet dimension romances of the likes of *Flatland* are not as widely available, and their relatively abstruse subject matter can make them less accessible now that the specific moment of intellectual curiosity about n-dimensional geometry has passed. Modern readers, for example, can empathize with the horror that Alice feels as she first begins to shrink, grow, and change form; fewer would, perhaps, empathize with the horror that the Square, the narrator of *Flatland*, feels when he watches the Sphere enter a locked two-dimensional room by moving through the third dimension.

But the influence of these works remains, not least in the space where biology and the fourth dimension overlap. Kreisel (2014: 399) notes that one of the core commonalities of dimensional romances is "an anxiety about the effects that higher-dimensional space would have on the integrity and privacy of three-dimensional bodies." Such stories, then, often have a strongly biological component. A key case in point is H.G. Wells' "The Plattner Story" (1897), in which a mild-mannered teacher is accidentally catapulted into the fourth dimension, where he can observe his neighbors when they believe themselves alone (as in Hinton's description, what are barriers in the third dimension are no barriers to a denizen of the fourth dimension). Plattner's adventures in the fourth dimension are spiritual and philosophical, but the consequences are biological: he returns to our dimension with the "unsymmetrical" parts of his body (liver, lungs, even his dominant hand) inverted (Wells 1897: 9–10). Plattner's travels in the fourth dimension are written on his body. The better the Victorians understood biology and the better they understood dimensionality, the more they saw how the one interacted with the other. Kreisel (2014: 408) argues that for writers of other-dimensional romances, "the gutting of the privacy and bodily integrity [...] allows a greater understanding of the nature of the universe as fundamentally complete": for the antagonists of the Twelfth Doctor episode "Flatline," this is literally true.

Dimensionality is central to the Doctor, whose mode of transport is, after all, Time and Relative Dimension in Space. Yet few stories explore dimensionality beyond the model seen in, for example, "Inferno" (1970) or "Rise of the Cybermen/The Age of Steel" (2006) and the related "Army of Ghosts/Doomsday" (2006). In these, the parallel dimension into which the TARDIS stumbles is, barring some historical and technological divergences, identical to our own, in that existing within it puts no physical or psychological strain on someone from our dimension, beyond that which results from meeting your own double. Episodes that show the vast spaces of the bigger-on-the-inside TARDIS—such as "The Invasion of Time" (1978), "The Doctor's Wife" (2011), and "Journey to the Centre of the TARDIS" (2013)—show similar limitations: they play with the ship's inherent characteristic of bigger-on-the-inside, but do not posit any particular strain on the TARDIS's inhabitants: where they do, as in "The Doctor's Wife," the stress is external (from the alien intervention of House) not internal (from the nature of the TARDIS). But occasionally, the program pushes the boundaries of dimensional existence. "The Celestial Toymaker" (1966) and "The Mind Robber" (1968), for example, are both set in places in which planes and dimensions are unreliable, at least to the TARDIS crew. Similarly, the null space between E-space and N-space in "Warriors' Gate" (1981) is a place of dimensional instability that poses a direct threat to the TARDIS and its crew, especially when the space gets smaller and smaller. And the Escher-ine spaces of "Castrovalva" (1982), created directly from the mathematical mind of companion Adric, confuse and disorientate the Doctor and his companions. But the show's most direct connection to its Victorian predecessors is also the most recent story examined here: "Flatline."

"Flatline" brings the Twelfth Doctor and companion Clara Oswald to Bristol, where the Doctor is trapped inside his own shrunken TARDIS, leaving Clara and graffiti artist Rigsy to prevent the incursion into our world of sinister two-dimensional beings that the Doctor calls the Boneless. The episode's most visceral images are the meeting of biology and dimensionality that result from the two-dimensional antagonists' experiments: humans reduced to outlines of their own nervous systems, spread out across walls like maps of the London Underground. Dimensionality is central to "Flatline," from the Doctor's description of the "dimensional leaching" that causes the TARDIS to shrink to his wry acknowledgment that the standard "It's bigger on the inside!" comment has never been truer. This is a story with no time travel, barring their arrival from the Orient Express in space to contemporary Bristol; rather, the "dimension" of Time and Relative Dimension in Space comes into its own.

The Boneless are attracted, first, to the impressions that humans make in two-dimensional space: handprints, tire tracks, footprints. Then, they draw the three-dimensional into two-dimensional space, something that can only be deduced from the flattened relics that remain: an abstract paint mark is an elongated human face, a "desert scene" painted on a wall is enlarged skin, a mural is a human nervous system, a graffiti memorial for the dead is instead the two-dimensional dead themselves. Finally, they occupy the shapes of the dead for an incursion into three-dimensional space, the bodies flickering in and out of coherence. They explore three dimensions as we, the three dimensional, explore two dimensions. The Doctor argues that a universe with only two dimensions has long been theorized. His implication is that it has been theorized by the Time Lords and their ilk ("dimensions are kind of our thing"), but, as we have seen, it was also theorized by the Victorians: the Boneless' ancestors are not *Doctor Who* villains such as the Daleks, but Victorian experiments such as the sentient Square of the Victorian *Flatland*.

What "Flatline" does not take from *Flatland*, however, is a sense in which a crossing of dimensions enables us not only to understand the alternative mode of being, but also to better understand our own. The Boneless never articulate their own intentions: they seem malicious, in their deliberate targeting of the community service workers by number, and the Doctor certainly interprets them as such. But it is unclear, for example, whether the two-dimensional relics are the product of malice or the simply the result of a three-dimensional body entering a two-dimensional space: that is, are the Boneless drawing people in to kill them, or are they drawing them in and thereby killing them? Whichever the impulse, the dead remain dead; no glimpse is given of the Boneless' two-dimensional world. The story becomes regressive. In this, it is a contrast to its Victorian predecessors. *Flatland*, for example, contains moments of existential horror akin to Clara and Rigsy's realization that the "mural" is a human body:

> An unspeakable horror seized me. There was a darkness; then a dizzying sensation of sight that was not like seeing; I saw a Line that was no Line; Space that was not Space; I was myself, and not myself. When I could find voice, I shrieked aloud in agony, "Either this is madness or it is Hell." "It is neither," calmly replied the voice of the Sphere, "it is Knowledge; it is Three Dimensions: open your eyes once again and try to look steadily" [Abbott 2002: 155].

In *Flatland*, the movement between dimensions is mutual (the Sphere visits two dimensions, the Square three) and the purpose educative: the Square learns not only about three dimensions, but also about the unspoken restrictions of his own space, in a commentary on Victorian social mores. In "Flatline," the movement between dimensions is one directional and the purpose aggressive: the protagonists learn nothing of the two-dimensional space (surprisingly, given that it has "long been theorized"; the Doctor, at least, might be assumed to have some curiosity). What commentary on social mores there is (in the abusive demeanor of the man in charge of the community service crew) is separate from the alien threat.

Nevertheless, "Flatline" shows the lingering effects of the concerns about bodily autonomy across dimensions that were raised by the late-Victorian scientific romances. The connection to *Flatland* suggested by the title is not coincidental: scriptwriter Jamie Mathieson notes *Flatland* as a key influence and the Abbott estate on which the story takes place is named for the author Edwin Abbott Abbott (Mathieson n.d.). Beyond this, the similarities between the Doctor and the imagined denizens of the Victorian fourth dimension are striking. Visceral, horrifying, if ultimately regressive, "Flatline" foregrounds dimension over time in a way that is uncommon in *Doctor Who*, and binds the modern text to its late Victorian predecessors.

The Archaeology of Empire

In the nineteenth century, archaeology as a science moved from the enthusiast to the scientist, from the private to the public, from cabinets of curiosities in wealthy households to museums in metropolitan centers, from "antiquities" to "archaeology." Victorian archaeology occupies a space between the curio hunters of the eighteenth century and the period in the twentieth century, when, as Howard Carter lamented, "tourism and popular culture were fomenting a facile curiosity […] that threatened to displace or prevent a more authentic relationship with the past" (Driscoll 2017: 109). The "fathers" of archaeology (Carter excluded) are figures we associate with the late eighteenth or (predominantly) the nineteenth centuries: Johann Joachim Winckelmann, whom C.W. Ceram lists first in *Gods, Graves, and Scholars*, observed the archaeological diggings at Pompeii and Herculaneum in 1758 and 1762, but Giovanni Battista Belzoni excavated Egyptian antiquities under the patronage of Henry Salt in 1815 to 1819, Paolo Emiliano Botta excavated the capital of Assyria in 1842 and 1843, and Austen Henry Layard excavated Nimrud and Nineveh from 1848, building on Botta's work.

The digs and the artifacts themselves are not the only significant flourishing of archaeology in the nineteenth century: what distinguishes Victorian archaeology from its predecessors is the theoretical growth of the field, the rise of the concept of "deep time." Although the term itself is best attributed to John McPhee in 1981 (Griffiths 2000), the concept of "deep time" emerged in the late eighteenth century with Scottish geologist James Hutton and was taken up enthusiastically across the nineteenth century. In 1866, Joseph Prestwich, reporting on the antiquity of flint tools in the *Proceedings of the Royal Institution of Great Britain*, noted that he "doubted whether, prior to 1858 and 1859, there were twenty men of science in Europe who would have admitted the possibility of the contemporaneity of man and of the extinct mammalia" (213). At the same time as the European empires were extending outwards across the globe, they were extending backwards into the spaces of the past. These conceptualizations of contemporaneous humanity and extinct species, of vast spaces of geological time and of human existence, underpin the development of time-traveling science fiction such as *Doctor Who*. The question is what else of the nature of Victorian archaeology such programs carry forward.

Tim Murray (1993: 177) argues that conflict in archaeological perspectives arises when "first practitioners, and subsequently members of the general public, have to find a way of comprehending information about the human past which has the potential to throw normative structures into disarray." One such normative scientific structure that Victorian archaeology threatened to disrupt was extant notions of race. Patrick Bratlinger (2011: 6–8) maintains that "racism informed virtually all aspects of Romantic and Victorian culture," and the "powerfully symbiotic" relationship between racism and imperialism means that "racism in its supposedly scientific forms was basic to the colonizers' mapping, census taking, legal and taxation systems, anthropology, and general understanding of the colonized." Just as this was true in the Victorian present, so it is true in their conception of the past. As I have suggested above, colonization did not only take place in the present: archaeological excavations often required or partook of colonizing practices, of which the struggle between the French and English for control of Egypt at the turn of the nineteenth century is perhaps the most dramatic example. The colonizing impulse and its associated "racial science" also carried back into the past: what John Reider (2012: 98) identifies as a Victorian paradigm of "accounts of savage societies that integrated them into a universal development theory" was also applied to the newly excavated civilizations. For example, the publicization of ancient ruins in Zimbabwe in the 1870s triggered the "racist view that the ruins must have been erected by Phoenicians or some other white race" (Bratlinger 2011: 169) rather than a rethinking of specifically Zimbabwean history.

As with dimensionality, the expansion of archaeological diggings and increased theorization about the human and pre-human past also triggered a new genre of fiction. In digging into *Doctor Who*'s concern with archaeology, we unearth the bones of authors such as H. Rider Haggard, who, in Patrick Bratlinger's (2011: 159) words, "penned adventure stories that helped set the pattern for fiction combining geographical with archaeological discovery." Leonard Driscoll (2017: 110) argues that Haggard's work was predicated on "[c]onservative, imperialist, and orientalist attitudes," but Allan Quatermain and other heroes in such novels as *King Solomon's Mines* (1885), *She* (1887), and *Allan Quatermain* (1887) provided the model for later archaeologist characters (and adjacent figures), including Indiana Jones and Lara Croft. And these works not only carried over the Victorian era's obsession with archaeology, they also retained archaeology's racist undertones. Bratlinger (2011: 169), for example, notes that Haggard "never relinquished his belief that a white race, probably the Phoenicians, had long ago colonized southeastern Africa, as Cecil Rhodes and the British were doing in the modern era"; his "lost races," such as the Zu-Vendi in *Allan Quatermain* or She in the titular novel, reflect this belief. But Haggard's particular version of Victorian archaeology remains influential, and its traces are evident in how archaeology appears in *Doctor Who*.

Archaeology has infused *Doctor Who* since the early historical stories of the original series: "Marco Polo" (1964), "The Aztecs" (1964), "The Romans" (1965), "The Myth Makers" (1965) and others all betray an archaeological obsession with re-creating and re-experiencing the past, often with the design of sets and props based on careful research in museums. Archaeological digs and other excavations act as inciting events (of greater or lesser significance) in both old and new series, including the finding of Eldrad's hand in "The Hand of Fear" (1976), the archaeological survey that releases Sutekh in "Pyramids of Mars" (1975), the human skull excavated in Kenya at the outset of "Image of the Fendahl" (1977), Professor Emilia Rumford and her study of the Boscombe Moor standing stones in "The Stones of Blood" (1978), Peri's archaeologist stepfather and the Trion artifact in "Planet of Fire" (1984), the excavation of the sword Excalibur in "Battlefield" (1989), the wreck of the spaceship *Byzantium* in "The Time of Angels/Flesh and Stone" (2010), the uncovering of the Dalek in "Resolution" (2019), and the recurring character of River Song, introduced as part of an archaeological survey in "Silence in the Library/Forest of the Dead" (2008). One of the most famous of all lost civilizations, Atlantis, makes appearances in "The Underwater Menace" (1967) and "The Time Monster" (1972). *Doctor Who* also has its own variants of lost civilizations, including the recurring figures of the Ice Warriors ("The Ice Warriors," 1967; "The Seeds of Death," 1969; "The Curse of Peladon," 1972; "The Monster of

Peladon," 1974; "Cold War," 2013; "Empress of Mars," 2017), the Sea Devils ("The Sea Devils," 1972; "Warriors of the Deep," 1984), and the Silurians ("The Silurians," 1970; "Warriors of the Deep," 1984; "The Hungry Earth/Cold Blood," 2010; and the recurring character Madame Vastra). These lost civilizations, as the airdates indicate, only emerged with the move away from the pure historical serials; here, the parallels with Victorian archaeology and archaeological romances sharpen.

One *Doctor Who* story draws strongly from both archaeology and lost civilization romances. Like many other lost civilization stories of the original series, "The Tomb of the Cybermen" is set in an unspecified future: it is not until the new series that *Doctor Who* explicitly repositions its lost civilization narratives in their nineteenth-century origins, in both Madame Vastra and "Empress of Mars." "Tomb of the Cybermen" finds the Doctor, Jamie, and new companion Victoria Waterfield on the planet Telos, where they find an archaeological expedition looking to excavate the legendary tombs of the Cybermen, who disappeared from the galaxy centuries ago. But while the archaeologists are motivated by their science, their financiers, who belong to the shadowy Brotherhood of Logicians, have their own agenda: to reanimate the Cybermen and seek their help in enabling the Brotherhood to rise to power. Kate Orman (2017) rightly argues that "The Tomb of the Cybermen" is a sibling story to "Pyramids of Mars" and "The Talons of Weng-Chiang" (1977). Like these, it draws from genre predecessors, such as Hammer mummy movies. But the mummies themselves are not the direct influence here, rather the means by which the mummies are uncovered: the archaeological surveys, their scientists, their financiers, and their imperialist trappings. For the latter, we look particularly to the figure of Toberman, one of the few characters in the original series played by an actor of color and a character type that Elizabeth Sandifer (2018: 244) identifies as "mute black strongmen."

The archaeologists are not the villains in "The Tomb of the Cybermen" but rather the financiers are: Kleig and Kaftan (and, by extension, Kaftan's "servant," Toberman). Ethnic origin is strongly demarcated in "Tomb": the archaeologists are English, the astronauts are American, and the logician-financiers, Kleig and Kaftan, are "European." Kleig, whose mildly Germanic name is more recognizably European than "Kaftan," is played by Greek-Cypriot actor George Pastell, who was known to audiences for playing vaguely "Eastern" villains in Hammer horror films. Similarly, Danny Nicol (2018: 52) describes Kaftan as "a continental European logician" and there are persistent anecdotes that actor Shirley Cooklin had her complexion darkened for the role.[1] As logicians, they are aligned with both the science of the Cybermen (whose logic puzzles they must solve to reanimate them) and the Doctor (whose superior logic ultimately solves the puzzles).

Toberman is aligned with the logicians but is not of them. Nor is he of the group of astronauts and archaeologists, nor of the Cybermen. Rather, he represents a third space, the space of Victorian archaeological romances.

The first the audience sees of Toberman is when a group of white archaeologists and astronauts bellow at him to keep his "big head" down as he appears silhouetted on a hillside. Toberman is a towering physical presence, a head taller than any other character, capable of dragging open the massive doors of the Cybermen's city that have already killed one man, but almost entirely silent: the actor, Roy Stewart, was formerly a stuntman and later reappears as a circus strongman in "Terror of the Autons" (1971), another "mute black strongman" role. Further, Toberman's strength is as much a weakness: he is not as strong as the Cybermen (the Cybercontroller at one point lifts him above his head and throws him across the room), but is sufficiently strong enough to attract their interest and become the first (and ultimately only) member of the party to be cybernetically converted. His strength ultimately causes his death. Danny Nicol, pointing to Stewart's roles as among the earliest non-white characters in the series, points out that "Brawn is snidely presented as an alternative to brains; and through silence, marginalised people are projected as mindless, voiceless and subject to the whims of external authority" (Nicol 2018: 52). More than that, "Tomb" positions Toberman as not only marginalized but as liminal.

The Cybermen, who first made their appearance in *Doctor Who* in 1966, may be one of the earliest examples of cyborgs on television (Geraghty 2008). That they are cyborgs and not robots is significant, because the cyborg draws physically, and not simply conceptually, on the human body (or, in this case, the Mondasian body) for its construction. The cyborg allows "The Tomb of the Cyberman" to posit two ways of being: human and inhuman, archaeologist and excavated, observer and subject. That the Doctor is outside this binary (by not only being not human, but also being aligned in logic with the Cybermen) is part of the nature of *Doctor Who*. That Toberman is outside this binary is more problematic. Toberman's liminal position in the archaeological survey recalls Umbopa in *King Solomon's Mines*. When Umbopa makes his first appearance, Quatermain describes him as "very light-coloured for a Zulu"; "his skin looked scarcely more than dark, except here and there where deep black scars marked old assegai wounds." Umbopa is, in his own words, "of the Zulu people, yet not of them." Similarly, Toberman is of the archaeological survey, yet not of them; after his partial conversion to Cyber-status, he is of the Cybermen, yet (as his saving of the survey demonstrates) not of them. Toberman is not only looking forward to the "mute black strongmen" of early television, he is looking back to the racial archaeology of the nineteenth century.

"The Tomb of the Cybermen" does not, at first, seem a Victorian episode of *Doctor Who*; it has no nineteenth-century setting or trappings. But it draws a direct line back to the nineteenth century with its archaeological romance of the lost tombs of a vanished race. The shades of Haggard and of the colonizing archaeological practices that created a space for his fiction would be evident in the story regardless. But the towering figure of Toberman throws them into sharp relief: silent, liminal, and self-sacrificing, Toberman draws the Victorian origins of stories such as "Tomb" out into the light for us to examine.

The Spirit of Collecting

In assessing nineteenth-century archaeological impulses, Tim Murray (1993: 183) (channeling Gillian Beer) suggests that

> The discovery of deep time, and of the possibility that memory would be overcome by oblivion [...] provided a threat to scientist and artist alike, a threat which was defused by effectively turning deep past (oblivion) into ethnographic present (living representations of stages of European race memory).

Murray (1993: 177) is talking specifically of the nineteenth-century creation of a plausible prehistory "which was broken up into a series of ethnographic presents linked vertically by small-scale processes such as diffusion and migration to explain demonstrable change." But his argument also intersects with the rise of museums in the nineteenth century. Museums were a means of bringing the deep past into the immediate present, and of doing so in a manner that turned human history in public space: just as the Victorians constructed the deep past as a series of ethnographic presents, they constructed the museum as a series of galleries that simultaneously contained and projected human history. Lara Kriegel (2006: 681) puts the matter even more bluntly:

> Museums are, unquestionably, among the most Victorian of institutions, monuments to their age and "master pattern[s]" of its cultural logics. Their edifices provide living tributes to the nineteenth-century projects of liberal reform, urban government, and imperial engagement. Their holdings offer lasting reminders of the Victorian preoccupations with collection and classification.

So here, in the final section of this essay, we come to the place where Victorian science gathers its artifacts.

Museums proliferated alongside and as a result of other popular outreaches of Victorian science: Amy Woodson-Boulton (2008) notes that by 1888, there were 211 provincial museums in Great Britain, and 100 of those had been built since 1872. The more the Victorian appetite for scientific enquiry grew, the more the museums grew, collecting the materials that shaped Victorian perceptions of the changing world in a flurry of "nineteenth-century European imperial kleptomania and historicism" (Woodson-Boulton 2008: 114). As the archaeologists dug up the empires of the past, they fed the resulting artifacts back to the empire of the present, which catalogued, displayed, and therefore retroactively contained them. Woodson-Boulton (2008: 112) argues that the origins of Victorian museums reveal a culture that was "deeply self-conscious about the very leisure time, wealth, and material abundance that allowed their production" and that the museums themselves make this self-consciousness manifest in five distinct ways: self-consciousness about the future; self-consciousness about history and time; self-consciousness about culture; self-consciousness about representation; and self-consciousness about the performance of identity. In this final section, I shall explore how the self-conscious Victorian museum is projected into the space of the 1960s.

Since archaeology is central to *Doctor Who* and recurs in greater and lesser ways throughout both original and new series, it is unsurprising that museums, too, make their appearance, although perhaps to a less prominent extent. In "The Seeds of Death" (1969) the TARDIS first materializes in a museum of rockets, which the new technology of T-Mat has seemingly rendered obsolete. In "Snakedance" (1983), the Manussan Empire exists both in the present (in the form of its returned enemy, the Mara) and the distant past (in precious archaeological artifacts). The museum of pacifism in "The Dominators" (1968), like the museum in "The Seeds of Death," houses objects that civilization has considered obsolete but is forced to reclaim (in this case, weapons). Museums also make briefer appearances in "The Keys of Marinus" (1964) and "The Web of Fear" (1968). The collection of planetary specimens in "Nightmare of Eden" (1979) straddles a line between a formal museum and a private collection, and other private collections play a role in "The Seeds of Doom" (1976), "City of Death" (1979), "Black Orchid" (1982) and "Ghost Light" (1989), as well as the P.T. Barnumesque circus collections in "Carnival of Monsters" (1973) and "The Greatest Show in the Galaxy" (1989). The new series also explores museum spaces, including the Vatican collections of "Extremis" (2017), the emotional coda to "Vincent and the Doctor" (2010) in the Musée d'Orsay, the museum's role in preserving and storing the Pandorica in "The Big Bang" (2010), and Henry van Statten's private museum of artifacts in "Dalek" (2005), which brings Daleks into the new series. But few of these stories interrogate the role, function, and history of the museum space to the extent evident in "The Space Museum." "Dalek," as befits an early episode of the new series, has an elegiac tone as the Doctor and Rose first explore the private museum, walking down rows of inanimate and mounted former enemies (Holdsworth 2011), although this is soon superseded by faster-paced adventure. "The Big

Bang" and "Vincent and the Doctor" construct museums as places in which the artifacts of the dead are stored, displayed, and venerated, but in both cases the space primarily serves as a vehicle for an emotional crisis. Amy Holdsworth (2011: 127) argues that the program as a whole is a museum, "a 'receptacle' for multiple forms of history, memory and identity." But only "The Space Museum," which takes place entirely in a museum, lingers on museum spaces as other than a plot device.

One of the last stories with original companions Ian and Barbara, who leave in the next story, "The Chase" (1965), "The Space Museum" takes place in the vast galleries of a museum to the lost glories of the militaristic Moroks. The Morok empire has since declined, and the Morok in charge of the museum is a colonial governor and administrator, petty and bored. The real work of the museum is done by the Xerons, the colonized inhabitants of the planet on which the museum is built, whose simmering rebellion is fomented by the TARDIS crew. It is also a story in which the museum's ability to show us ourselves is literalized, as the TARDIS crew come face to face with themselves as museum exhibits. "The Space Museum" predates the other stories discussed in this essay, but viewed from two decades deep in the twenty-first century, it seems a distillation of them: the time travelers existing in more than one dimension, declining civilizations, colonialism, lost glories—a fitting story to end this analysis of *Doctor Who*'s relationship with Victorian scientific practices.

Just as with Victorian museums, per Woodson-Boulton, "The Space Museum" exists in a state of self-consciousness. From the outset, the characters are surprised to find that space has a museum, and indeed the Smithsonian's National Air and Space Museum had only been established in 1946, as the National Air Museum: the "space" component was integrated during the space race. Despite this, the TARDIS inhabitants respond as they would to any museum, with Vicki telling Ian that she feels they should be shadowed by a guard as they walk through the galleries. Since Vicki is from the twenty-fifth century, this argues either for a certain, rather implausible, degree of continuity in museum management or for a consciously performative approach to prescribed museum behavior on the story's part. Similarly, when the Doctor discourages Vicki from touching the exhibits, she finds that she is unable to do so, because she and they do not occupy the same dimension. Finally, towards the end of the first episode, the TARDIS crew come face to face with themselves as exhibits, frozen in glass cases after having jumped a time track. The story thereafter exists in tension between the present and the future, as the crew try to avoid a fate they have already seen. "The Space Museum" draws on audience expectations of behavior in the Victorian edifices of the museum in order to deliver its uncanniness about time, about the future, about representation, about the distinction between "us as museum visitors" and "us as museum exhibits."

The same tension and uncanniness exists on a broader scale and, again, the bones of Victorian museum building lie beneath. The Moroks turn a colonized planet into an enormous museum to their past conquests, but the museum is devoid of any visitors but the TARDIS crew: "People tire of their heritage," the governor, Lobos, explains to the Doctor. This museum is an inversion of the British museums with which the audience would be familiar and which burgeoned in number in the nineteenth century: there, the museums sit at the heart of empire, and the treasures of colonial spaces are sent back to fill them. The planet itself turns against the building, which, complete with its contents, is slowly decaying in a poisonous atmosphere. One of the intriguing elements of this story, which is absent from other *Doctor Who* stories that showcase museums as locations, is the extent to which this story shows the curatorial aspect of collection building. In the poisonous atmosphere, the Moroks perform running repairs on decaying exhibits, despite the lack of attendance: the audience sees Morok bureaucrats discussing how much longer exhibits will survive and companion Barbara hiding in a storage space filled with items rotated out of the galleries. As the TARDIS crew walk through endless galleries of rifled imperial trophies, poisoned slowly by the atmosphere (and at one point poisoned rapidly by the Moroks), they are also walking through the history of museum building, back through the galleries of countless Victorian museums that still sleep beneath the bones of even a Space Museum. What was imagined as a "lasting memorial to the achievements of the Morok Empire" is repeatedly presented as merely temporary, much like the derelict museum called the Palace of Green Porcelain in Wells' *The Time Machine* (1895) or the legs of Ozymandias in Shelley's 1818 poem. If, as Kriegel argues, Victorians imagined museums as "monuments to their age and 'master pattern[s]' of its cultural logics," "The Space Museum," more critical in its relationship to its Victorian roots than perhaps "The Tomb of the Cybermen," suggests that such monuments are both predatory and transient. Even the great museums of empire change or fall—or, at the very least, their attendances do.

We do not need the concept of deep time to recognize that the nineteenth century is not so far removed from *Doctor Who*, particularly in its earlier stories. Whether the episodes take place in the nineteenth century or not, the scientific preoccupations that shaped Victorian culture and Victorian popular fiction reverberate through *Doctor Who*, from the earliest seasons to the most recent regenerations, from the spaces of museums to the tombs of dead races to the dimensions that lie beyond our own. Indeed, where the nineteenth-century setting is most distant is sometimes where the influences are most apparent. In "Flatline," the program's ongoing fascination with dimensionality is stripped back to the mathematical puzzles and existential horror of Victorian incursions into hyperspace. In "The Tomb of the Cybermen," the undercurrents of race science show us that archaeologists of the distant future are not so far removed from their nineteenth-century predecessors. And in "The Space Museum," the program dwells on the vast galleries of Victorian museum building and their predatory undercurrents. The First Doctor, wandering through the Space Museum in waistcoat and watch chain, gives way to the Second Doctor among the Tombs of the Cybermen, and the Second Doctor becomes the Twelfth Doctor in contemporary Bristol confronting enemies from another dimension. From overt references to buried patterns of understanding, the nineteenth century always sleeps in the mind of the show.

Note

1. I have not been able to isolate the original of this anecdote, which is presented as common knowledge in a number of blog posts on this story.

References

Abbott, E.A. ([1884] 2002) *The Annotated Flatland.* New York: Basic Books.
Bratlinger, P. (2011) *Taming Cannibals: Race and the Victorians.* Ithaca: Cornell University Press.
Ceram, C.W. (1952) *Gods, Graves and Scholars: The Story of Archaeology.* London: Victor Gollancz.
Driscoll, L. (2017) Restoring the lost empire: Egyptian archaeology and imperial nostalgia in H. Rider Haggard's "Smith and the Pharaohs" (1912). *Nordic Journal of English Studies* 16(2): 108–128.
Geraghty, L. (2008) From balaclavas to jumpsuits: The multiple histories and identities of Doctor Who's Cybermen. *Atlantis* 30.1: 85–100.
Griffiths, T. (2000) Social history and deep time. *Tasmanian Historical Studies* 7(1): 21–38.
Haggard, H.R. (1885) *King Solomon's Mines.* United Kingdom: Cassell and Company.
Hinton, C.H. (1897) *What Is the Fourth Dimension?* London: Swan, Sonnenschein & Co.
Holdsworth, A. (2011) *Television, Memory and Nostalgia.* London: Palgrave Macmillan.
Kreisel, D.K. (2014) The discreet charm of abstraction: Hyperspace worlds and Victorian geometry. *Victorian Studies* 56(3): 398–410.
Kriegel, L. (2006) After the exhibitionary complex: Museum histories and the future of the Victorian past. *Victorian Studies* 48(4): 681–704.
Mathieson, J. (n.d.) The Boneless. *Jamie Mathieson: Screenwriter.* www.jamiemathieson.com/the-boneless (accessed 16 November 2019).
Mills, C. (2013) "Such a dazzling display of lustrous legerdemain": Representing Victorian theatricality in Doctor Who. *Neo-Victorian Studies* 6(1): 148–179.
Murray, T. (1993) Archaeology and the threat of the past: Sir Henry Rider Haggard and the acquisition of time. *World Archaeology* 25(2): 175–186.
Nicol, D. (2018) *Doctor Who: A British Alien?* London: Palgrave Macmillan.
Orman, K. (2017) *Pyramids of Mars. The Black Archive 12.* Edinburgh: Obverse Books.
Prestwich, J. (1866) On the quaternary flint implements at Abbeville, Amiens, Hoxne, etc., their geological position and history. *Proceedings of the Royal Institution of Great Britain* 4: 213–222.

Reider, J. (2012) *Colonialism and the Emergence of Science Fiction*. Middletown, Wesleyan University Press.
Sandifer, E. (2018) *Tardis Eruditorum Volume 2: Patrick Troughton*, 2nd edition. United States of America: Eruditorum Press.
Stewart, I. (2002) Introduction. In: Stewart, I. (ed.) *The Annotated Flatland*. New York: Basic Books, pp. xiii-xxvii.
Throesch, E. (2009) Nonsense in the fourth dimension of literature: Hyperspace philosophy, the "new" mathematics, and the *Alice* books. In: Hollingsworth, C. (ed.) *Alice Beyond Wonderland: Essays for the Twenty-First Century*. Iowa City: University of Iowa Press, pp. 37-52.
Wells, H.G. (1895) *The Time Machine*. United Kingdom: William Heinemann.
Wells, H.G. (1897) The Plattner story. In: *The Plattner Story and Others*. London: Methuen & Co., pp. 2-28.
Woodson-Boulton, A. (2008) Victorian museums and Victorian society. *History Compass* 6(1): 109-146.

Doctor Who and the Dinosaurs

Spectacle, Monstrosity, Melodrama and Ideology in Dinosaur Mediations

Ross Garner

Introduction

Doctor Who's relationship with representing the concept of "deep time" arguably begins with the first televised story, "An Unearthly Child" (1963), due to its depiction of Neolithic cave dwellers. However, it was not until "The Silurians" (1970) that the series attempted to engage with the Mesozoic period, that is the "'middle animal' era, containing as it did many bizarre and gigantic saurians" (Noble 2016: 46–7), at either a denotative or connotative level. James Chapman (2013: 85) partly attributes this story's significance to its coding of the narrative threat as something internal "within the Earth" rather than outer space. Much like other stories representing the Mesozoic, though, the dominant interpretation of "The Silurians" has been in terms of its socio-historical meanings in the context of 1970s British politics. Writing on this story and its stablemate, "The Sea Devils" (1972), Chapman (2013: 86) argues that "the belligerent factions amongst the Silurians and Sea Devils stand for the PLO [Palestine Liberation Organization] and the IRA [Irish Republican Army]: groups claiming to represent the dispossessed and displaced inhabitants who resort to violence to assert their territorial claims." Similarly, "Invasion of the Dinosaurs" (1974) has primarily been aligned with "the emergence of the eco-catastrophe narrative" (Chapman 2013: 89) in the series as a result of the formation of ecologically-minded protest groups such as Greenpeace. On the one hand, these established accounts point to authorially-focused interpretations as each was written by screenwriter Malcolm Hulke. On the other, the recurrence of these readings across both academic and fan interpretive communities (Fish 1980) could be assigned to the presumed (lack of) aesthetic quality which characterizes the realization of prehistoric creatures in these stories; as Chapman (2013: 91) summarizes, "'Invasion of the Dinosaurs' is often derided for its unconvincing puppet dinosaurs."

One consequence of uncritically accepting these evaluations is the lack of serious discussion of how *Doctor Who*, as a popular long-running science fiction program, has contributed to constructing and communicating ideas concerning dinosaurs and the Mesozoic. This absence resonates with Brett Mills' (2017: 32) argument that "television's primary organising principle is one that cannot help but be anthropocentric" and as a consequence, serious analysis of how the medium engages with non-human representations more widely has been overlooked. Yet, while there have been recent attempts to decenter anthropocentricism by analyzing animals as either television representations (Mills 2017) or within science fiction as a genre (Vint 2010), these considerations are yet to fully extend to the concept of dinosaurs. This essay therefore provides an original intervention into the ongoing academic analysis of both *Doctor Who* and television's mediation of natural history by taking the program's construction of dinosaurs seriously, to address the continuities, differences and significance of how *Doctor Who* has constructed dinosaurs as meaningful.

It might, of course, be tempting to subsume *Doctor Who*'s representations of dinosaurs to more general discussions of monstrosity in the series (Mesozoic-coded or otherwise; see Sleight 2012: 74–81). However, such a position is unsophisticated as it ignores how dinosaurian creatures represent a different category to the Daleks and the Cybermen, for example. José Luis Sanz (2002: 47) rightly argues that dinosaurs "straddle the imaginary and real worlds" and as we "can be certain that the dinosaurs became extinct millions of years ago" they occupy a different cultural status to creatures of pure imagination. Guided by this understanding, the present essay focuses exclusively on instances when *Doctor Who* has explicitly depicted "giant lizards" in stories like "The Silurians" and "Invasion of the Dinosaurs" and, more recently, "Dinosaurs on a Spaceship" (2012) and "Deep Breath" (2014). However, it also extends the category of "giant lizards" to those connoting this status such as the Skarasen of "Terror of the Zygons" (1975) and the Myrka in "Warriors of the Deep" (1984). Although connotative dinosaurs might be more easily classifiable as examples of "science fiction monsters" due to their narrative coding as organic-inorganic hybrids (e.g., the Skarasen's cyborg status, the Myrka's ability to attack via electric charge), the essay instead demonstrates that *Doctor Who*'s connotative dinosaurs further reinforce established ways of understanding the dinosaur in the series. One consequence of focusing solely on dinosaurs as "giant lizards," however, is that stories featuring Mesozoic-coded humanoid monsters such as "The Sea Devils" and "The Hungry Earth/Cold Blood" (2010) are overlooked as neither includes such a creature at either the denotative or connotative level. The following arguments are subsequently drawn from textual analysis of twenty-three television episodes of "classic" and "new" *Doctor Who* comprising approximately 660 minutes (11 hours) of broadcast material.

Overall, the essay develops two arguments. The first is that dinosaurs in *Doctor Who* have primarily been constructed for dramatic purposes rather than as objects of scientific curiosity. However, while the show has not closed down the latter readings it has also not reduced connotative or denotative dinosaurs to the role of inhuman threat. Instead, the essay argues, through analysis of the show's narrative and aesthetic codings, that dinosaurs in *Doctor Who* have regularly been overdetermined, by being readable through intersecting conventions of dramatic spectacle, monstrosity and melodrama as well as scientific observation. The second argument concerns the significance of these findings within wider scientific and cultural discourses that position dinosaurs. Building upon Brian Noble's (2016) anthropological arguments concerning the need to see "scientific" and "fictional" constructions of the dinosaur as intertwined, I argue that *Doctor Who*'s overdetermined construction of dinosaurs challenges deep-rooted ideological meanings of dinosaurs by favoring a pro-biodiversity and environmentally-liberal perspective. I start from a position of accepting that "the scholarly literature on dinosaurs as public/scientific creatures [lacks] detailed case studies" (Noble 2016: 19) of how particular media properties and franchises have approached the dinosaur as a cultural, historical and scientific object. This essay responds to this absence by arguing that dinosaurs in *Doctor Who* have continually upheld multiple generic codings and that these, alongside other contextual factors, must be addressed when analyzing how the series mediates specific objects or topics of scientific interest.

Specimen, Spectacle or Both?

Writing in relation to the procedural crime drama *CSI: Crime Scene Investigation* (CBS 2000–15), Sue Turnbull (2007: 30) argues that "*CSI* allows us to 'look' [...] at the body" that is under investigation and that this opportunity to pay witness alongside the series' forensic scientist characters provided the show's "primary fascination" for audiences. Underpinning Turnbull's argument is, as Deborah Jermyn (2007) elucidates, the historical correlation between a medical-scientific gaze and detached, sustained observation that allows for objective knowledge about the subject under scrutiny to be obtained. This ocular-centric scrutinizing discourse is linked to Enlightenment thinking and the intersecting forms of institutional power that legitimize and sustain science and medicine as authoritative institutions for regulating knowledge of anatomy and illness (Foucault 1989). However, the "truth claims" (Jermyn 2007: 80) offered by popular fictional television are more complex than simply displaying or demonstrating scientific evidence to audiences. While primarily factual genres may "seem to express a particular kind of claim to reality and are presented as if the 'truth' of the events can be read or directly interpreted from the images themselves" (Lury 2005: 18), this is clearly not the case with fiction. As Karen Lury (2005: 47) states of *CSI*, "none of the potentially demonstrative images employed—X-rays, film footage, photographs and microscope slides—are in fact documentary images, as they have all been carefully fabricated for the series." Images purporting to "truthfully" demonstrate either scientific practice or the effect of biological reactions upon the human body in *CSI* are instead best understood as "embracing the 'realism' and the spectacle" (Jermyn 2007: 89) associated with mediating science to non-specialist audiences. Cumulatively, these debates allude to a deep-rooted cultural tension where, on the one hand, scientific information is expected to be communicated by a popular audio-visual medium like television in a manner that connotes objectivity. On the other, this expectancy sits alongside the need for fictional series like *CSI*, or *Doctor Who* in this instance, to dramatize processes, methods and objects of scientific enquiry in a way that attracts and maintains audiences.

Stories featuring both denotative and connotative dinosaurs throughout *Doctor Who*'s history have used the detached and objective scientific gaze for dramatizing scientific research. Episode 6 of "The Silurians," for example, features many scenes where the Third Doctor undertakes scientific procedure to attempt to find the antidote to the virus that has been released on the public via the infected Major Baker. Alternatively, the second episode of "Invasion of the Dinosaurs" includes a five-second static long shot of the Doctor observing a stegosaurus that has unexpectedly appeared in London. The Doctor's dialogue of "No, no. Don't shoot. I want to take a good look at it" positions the creature as an object for scientific observation and so (re)affirms the show's construction of the Doctor as a deductive inquisitor during this period (Newman 2005: 27; Chapman 2013: 78) . However, the Doctor's preceding line of "Good grief! It's a stegosaurus" implies that the creature's appearance should be read both diegetically and extra-diegetically (e.g., by the viewing audience) as a moment "of spectacle, [...] sometimes no longer than a shot, in which we are asked to *look*" (Wheatley 2016: 17; original emphasis) with the intention of inspiring awe and appreciation amongst audiences.

That this tension between observational gazing and dramatic spectacle is identifiable in this scene (and others where dinosaurs are encountered) suggests how *Doctor Who* has employed what Noble (2016: 13) calls "the specimen-spectacle complex" pertaining to prehistoric creatures. Developed primarily to account for exhibiting dinosaur skeletons in museums, the specimen-spectacle complex, Noble (2016: 16) argues, is an ongoing component of how meaning is given to prehistoric creatures where "Dinosaurs were an entity both constituted as specimen and, in turn, reconstitutable in public form as spectacle, and indeed as characters in science fiction literature and film." However, as the next section demonstrates, *Doctor Who* has arguably foregrounded the spectacular side of this complex throughout its history alongside aligning the creatures with additional generic codings.

The Overdetermined Dinosaur

Correlations between how *Doctor Who*'s classic series constructed and visualized dinosaurs and moments of spectacle are evidenced through the creatures' appearances as either early narrative hooks or at episodic cliffhangers. The former point is best exemplified by "The Silurians" where the appearance of a bipedal dinosaur (assumed to be a *Tyrannosaurus rex*) threatening two cavers provides the destabilizing narrative event.[1] The dinosaur's unexpected intrusion into the "everyday" setting spectacularly disrupts the status quo and sets into motion the story's opposition between representations of humanity and the (in this instance, Mesozoic-coded) Other that characterizes both narratives of human-dinosaurian encounter (Noble 2016) and science fiction plots more generally (Tellotte 2001). More common, however, is the latter narrative strategy where sudden encounters between dinosaurs and humanity provide the moment of suspense that is deployed with the intention of attracting a returning audience to the story's following episode (Dunleavy 2009). Four of the six episodes of "Invasion of the Dinosaurs" employ dinosaurs for these purposes, as does the second episode of "Terror of the Zygons." Similarly, episode 2 of "Warriors of the Deep" concludes with the Fifth Doctor and Tegan being threatened by the Myrka for the first time. That each of these shots conclude with the electronic sting of the series' theme music being employed on the audio codes dinosaurs as "strange and unknowable" (Wheatley 2016: 107) creatures and therefore as objects of spectacle.

One notable difference between how new *Who* has employed moments of dinosaurian spectacle and its classic incarnation concerns the shift to aggressively foregrounding rather than concealing prehistoric creatures.

Whereas "Invasion of the Dinosaurs" named its first episode "Invasion" to maximize the impact of dinosaurs appearing in the story, "Dinosaurs on a Spaceship" makes the presence of the creatures explicit within the episode title. Alternatively, the appearance of a T-rex in "Deep Breath" featured prominently in both trailers for the 2014 season and the Twelfth Doctor's debut episode. This shift can be aligned with industrial developments, namely the increasingly competitive environment which the program must negotiate. Helen Wheatley (2016: 9) argues "television has […] turned towards the spectacular at times of increased competition." Spectacular imagery on television, such as dinosaurs in this instance, should not be solely associated with the fragmenting audiences and increased choice offered by historical developments linked to digitization, as Wheatley (2016) skillfully demonstrates. However, the shift to employing prehistoric creatures for pre-publicity purposes in new *Who* indicates one of the ways the program has changed how it aligns dinosaurs with the category of spectacle. That a crashed model of the TARDIS atop of a T-rex skull was placed in Parliament Square, London to further publicize "Deep Breath" and the 2014 season underscores these associations.

Addressing industrial and institutional contexts further evidences how contemporary *Doctor Who* has used dinosaurs for commercial purposes such as building hype and attracting audiences to the series—both domestically and abroad. Within the UK, *Doctor Who* is expected to fulfill the role of "'consensus' audience grabber" (Hills 2010: 211) for BBC One by attracting a high audience share composed of multiple niche profiles (also Johnson 2005). Noble (2016: 6) notes that dinosaurs remain highly popular with "millions of European, Japanese, and North American children under the age of eight." Foregrounding the presence of dinosaurs within individual episodes thus works partly to target child audiences and attract these to twenty-first century *Doctor Who* with the intention of building a family audience while also assisting with overseas sales. Furthermore, making references to dinosaurs in episode titles and trailers is indicative of the series employing what Matt Hills (2010: 220; original emphasis) names "'*high concept*' TV" where a combination of "a one-line marketable hook" and "the use of pre-sold properties and highly recognizable 'generic icons' […] in altered contexts" assist in branding and marketing the show. Explicit mentions of dinosaurs fulfill this function while the continual suggestion of the creatures being temporally displaced suggests an original and intriguing premise that distinguishes *Doctor Who* from its competitors—telefantasy or otherwise.

Further evidence of dinosaurs in *Doctor Who* being contained within a spectacular, dramatizing discourse can be provided through analyzing different textual elements used for constructing the creatures. Appearances of dinosaurs in both classic and new *Who* consistently emphasize the creatures' size in comparison to human characters. One way this is highlighted is via dialogue: the Third Doctor remarks that the Silurians' cave-dwelling creature is "very, very large and very, very alive"; the Fourth Doctor states that the Skarasen is "a monster of terrifying size and power"; Jenny Flint remarks "Big Fellow, isn't he?" of the T-rex in "Deep Breath" while the Twelfth Doctor (erroneously) refers to the creature as "Big man." Such comments reinforce how since "animals don't look like humans [...] in visual terms they will always be alien, categorised as non-human" (Mills 2017: 47). Aesthetic codes of camera positioning, angle and framing then further accentuate the enormity of prehistoric creatures in relation to both human characters and/or recognizable (English) landmarks like the Westminster clock tower ("Deep Breath"). These strategies include having the head and neck of the dinosaur shot in close-up to emphasize its size in the frame (see "The Silurians," "Terror of the Zygons" or "Warriors of the Deep"), having the creature framed so that only a fragment of it is seen ("Terror of the Zygons," "Warriors of the Deep"), or, alternatively, shooting the human characters from the perspective of the dinosaur to demonstrate their minuteness ("Deep Breath").

These aesthetic codings might appear unsurprising as they affirm dominant meanings regarding prehistoric creatures. W.J.T. Mitchell (1998: 69) argues that "the meaning of dinosaurs and the reasons for their popularity" is frequently accounted for in terms of the creatures' "bigness, ferocity, rarity, antiquity, or strangeness." By foregrounding dinosaurs' size and scale, *Doctor Who* has therefore regularly supported how dinosaurs are "widely used as spectacles [...] in science fiction" (Noble 2016: 17) and so serve primarily dramatic purposes. However, I would argue that the aesthetic strategies employed also intersect with dinosaurs' meaning as objects of horror and monstrosity. José Luis Sanz (2002: 47) argues that "The enormous dimensions of a brachiosaur or a tyrannosaur can cause a feeling of uneasiness in human beings, who are fascinated by their potential for power and domination." Sanz's position is indebted to Noël Carroll's (1990: 17) theory of art-horror as a distinct mediated mode "in which the emotive responses of the audience, ideally, run parallel to the emotions of characters." In other words, as the audience's point(s) of identification in the narrative react negatively to encountering such creatures by evaluating the dinosaurs as both real and threatening, audiences are cued to respond in a similar manner.

There are multiple other ways that dinosaurs in *Doctor Who* can be discussed in relation to theories of monstrosity and art-horror. One approach would be to provide detailed analysis of the performance codes used by actors to react to prehistoric creatures, therefore advancing debates concerning television acting (see Caughie 2014; Lacey 2016). Cumulatively, these sequences suggest that "the appropriate reactions [comprise] shrinking, paralysis, screaming, and revulsion" (Carroll 1990: 18) when face-to-face with Mesozoic predecessors. This trope runs from the Third Doctor's reaction to encountering a bipedal dinosaur in the cave system at the end of episode 1 of "The Silurians" through to how the Eleventh Doctor, Rory and Brian protect themselves from a flock of pterodactyls in "Dinosaurs on a Spaceship" by retreating, recoiling and reducing their size to avoid the airborne predators. Studying performance codes could therefore assist in understanding how art-horror is performed as an acting style on television through mediations of dinosaurs.

At the same time, dinosaurs have been contained within discourses of monstrosity throughout *Doctor Who* through being encoded so that the protagonists "desire to avoid the touch" (Carroll 1990: 27) of the creatures since they represent an immediate threat to survival. The Myrka best demonstrates this point as the soldier grunts of Sea Base Four and Doctor Solow are immediately electrocuted upon making physical contact with the pseudo-reptilian creature. Alternatively, the Fourth Doctor avoids the slashing claws of the Skarasen ("Terror of the Zygons," episode 3) while Amy, Riddell and Queen Nefertiti protect their corporeality from a pack of velociraptors by using stun guns ("Dinosaurs on a Spaceship"). Recurrent across these constructions are associations between dinosaurs and discourses of impurity (Carroll 1990) and further analysis of this relationship could consider how and why representations of specific elements of prehistoric creatures such as claws and teeth are continually coded as foul.

Yet, alongside constructions of dinosaurs throughout *Doctor Who*'s history being readable as simultaneously spectacular and horrifying, the aesthetic encoding of the creatures also intersects with discourses delimiting melodrama. This point can partly be demonstrated by the creatures' use in cliffhangers in the classic series where the camera lingers on an image of a dinosaur for between four and ten seconds. On the one hand, these shots speak to the "imbalance between narrative and spectacle" (Wheatley 2016: 8) where the latter temporarily subsumes the former to simultaneously enhance the spectacle and threat of the dinosaur onscreen. On the other, the shot length arguably creates a tableau-esque effect that pauses narrative development and invites audiences to inspect the visualized creature onscreen. This technique is used throughout "Invasion of the Dinosaurs" as well as at the end of episode 1 of "The Silurians" and the Fourth Doctor's encounter with the Skarasen (see episodes 2 and 3 of "Terror of the Zygons"). More recently, it was also employed for the reveal of the ankylosauruses on the Silurian ark during the pre-credits sequence for "Dinosaurs on a Spaceship"—albeit in a manner where the melodramatic associations are reduced, as I'll return to shortly.

Roberta Pearson (1992: 39) argues that interrupting narrative progression to strike a static tableau is a convention originating in theatrical melodrama which is typically used "to convey intense emotions in nonverbal form, freezing in place with arms fully extended outward, downward, or upward at an act's climax." Although climactic sequences of dinosaurian display in *Doctor Who* demonstrate stasis by using a locked camera position to convey a heightened sense of wonder and/or fear, a completely frozen tableau is not offered. Instead, movement, however limited, occurs within the frame via the dinosaur prop or effect. The models in "Invasion of the Dinosaurs" have either moving heads or necks, the Skarasen puppet shifts its outward gaze across the frame, the Myrka's head bobs up and down, while the ankylosauruses boisterously shove each other as if competing for space. From one perspective, these developments demonstrate the advance of special effects for visualizing dinosaurs in popular tele-fictions by the increased opportunities for "realistic" depictions of movement that technological developments have offered. Alternatively, I would argue that the combination of the creatures' framing in either medium close-up or close-up, the lingering shot length, and the deployment of histrionic performance codes by actors responding to the creatures (Pearson 1992) combine to create tableau-esque moments that give rise to melodramatic connotations.

In sequences where the heavily artificial nature of the dinosaur props—especially with the historical examples where movement is limited—is accentuated, the possibility that audiences would locate the depicted creatures within a non-realist discourse also arises. Marcie Frank (2013: 538) argues that historicized definitions of melodrama have "something important to tell us about changes in the standards of realism" in relation to drama as the former is typically positioned in contrast to the latter and evaluated as inferior. If applied to the dinosaur models in *Doctor Who*, the shot length used for visualizing the models allows audiences to inspect these in detail, recognize the multiple signs connoting their artifice, evaluate them as excessive, and interpret them through a discourse of melodrama.

In comparison to the limited movements or denoted artifice of dinosaurs in classic *Who*, the jostling ankylosauruses in "Dinosaurs on a Spaceship" or the depiction of the T-rex in "Deep Breath"—where textures of skin and teeth are evident in the creature's visual design—are unsurprising. The movements of the creatures attempt to negate melodramatic associations by combining assumed photo-realist depictions with connotations of independent and verisimilar movements (see Pearson 1992).

These expectations have proved popular with audiences and are intended to permit the onscreen effect or prop to be read as more "realistic" than, say, stop-motion model work (Noble 2016). New *Who*'s dinosaurs are therefore visualized in terms of the high-cost special effects which assist in securing the show's ongoing status as contemporary "quality" television (Brunsdon 1997; also Hills 2010). The noise and threat coded into these dinosaurs do undoubtedly imply the "televisual 'whistles and bells'" (Wheatley 2016: 103) that are at odds with the upper-middle-class-based correlation between "quality," expense and visual spectacle that Charlotte Brunsdon (1997) has historicized. However, the fleeting appearances of the dinosaurs during these sequences arguably re-inscribes culturally acceptable notions of restraint by keeping the "whistles and bells" to a minimum.

Despite the "quality" aesthetic codings of dinosaurs in new *Who*, melodramatic readings of these creatures remain possible due to the purposes they serve within the narrative as both the T-rex in "Deep Breath" and the dinosaurs hosted within the titular spaceship are positioned as victims. The T-rex is burnt to death shortly after the Twelfth Doctor draws parallels between their temporally displaced status, while Solomon views the menagerie aboard the Silurian ark solely as commodities to sell. When Tricey the Triceratops is killed by Solomon's robots, the dinosaurs' encoding as innocent is further underlined (despite the episode's ongoing coding of species like velociraptor as threatening). Writing on animal (but alas not dinosaur) representations on television, Mills (2017: 4) notes that affective responses to imperiled creatures "relies on their individualised nature" within the context of specific episodes or programs. However, I would argue that Mills' position does not go far enough as it is by locating the construction of individualized creatures *within a discourse of melodrama* that accentuates an affective response for dinosaurs in *Doctor Who*. If it is accepted that part of the melodramatic mode involves "imperiled virtue, black villainy, and fantastic coincidences that allowed the former to triumph over the latter" (Pearson 1992: 8) then dinosaurs in new *Who* correspond to these meanings. The dinosaurs are narratively constructed as becoming caught up in human-alien confrontations and so suffer blamelessly at the hands of episodic antagonists before being either avenged, redeemed or saved by the Doctor. Thus, while new *Who*'s dinosaurs are less immediately aligned with aesthetic codes of melodrama than their classic equivalents, these associations have shifted to the narrative level. Throughout *Doctor Who*'s history, then, dinosaurs are been overdetermined as spectacular, simultaneously monstrous and melodramatic objects which primarily serve dramatic purposes at the expense of scientific inquiry.

The Ideological Dinosaur

What is the significance of breaking down the intersecting generic, narrative and aesthetic encodings that have continued to construct both denotative and connotative dinosaurs in *Doctor Who* as overdetermined? One immediate point of reference might be to use the previous section's arguments to critique the program's commissioning institution for failing to fulfill its public service obligations—either historically or at present. At its inception, the BBC's role was outlined as "serv[ing] its audiences as citizens to be educated and informed as well as consumers to be entertained" (Petley 2006: 42). Despite currently operating in a much-changed technological, organizational and competitive marketplace (Hendy 2013), similar institutional priorities remain identifiable. Currently, the Royal Charter (DCMS 2016: 3) states of the Corporation's Public Purposes that "the BBC should help everyone learn about different subjects in ways they will find accessible, engaging, inspiring and challenging." This Purpose extends to the BBC's drama output as the 2019–20 Annual Report identifies "learn[ing] about a part of history" (BBC Studios 2019: 20) as one requirement of fictional programs. The extent to which this responsibility extends to knowledge about Mesozoic "deep time" is, of course, open to debate—especially given the BBC's noted historical association with producing "quality" period dramas depicting Britain's imperial and aristocratic heritage to global audiences (Cardwell 2002). Nevertheless, *Doctor Who*'s continual commitment to representing dinosaurs through a combination of spectacular, horrific and melodramatic codes might be taken as at best constituting a weak commitment to providing audiences with scientific-observational, and therefore educational, mediations.

Such a critique would nevertheless produce the kind of knowledge blockage that Brian Noble (2016) has sought to complicate with regard to how the Mesozoic is made meaningful across cultural sites and institutions. Noble (2016: 34–5) challenges deep-rooted assumptions concerning dinosaurs in popular media which forcefully posit that "What are reckoned as sites for fictional accounts, as in the case of science fiction literature and cinema, are often engaged by scientists as locales to assert influence over scientific debates." That is, constructions of dinosaurs throughout the history of *Doctor Who* gain the attention of scientists for the purpose of pointing out their shortcomings in the communication of paleontological knowledge to lay audiences, but by offering these criticisms, the dominance of scientific discourse over TV fiction is (re)asserted. In contrast, Noble (2016: 35) draws upon anthropological and cultural perspectives

> to advance the proposition [...] that what are taken as media and literary performances or phantasies are also, quite properly, part of the *performativity* of palaeobiological

practice. They are part of the reality-making work of science, not something outside it, nor do they simply influence or bias it.

Noble (2016: 18) therefore argues for adopting a "both/and" approach to analyzing dinosaur mediations which engages the spectrum of factual and fictional depictions "to consider the actions on their own and together, the outputs of each, the sourcing for each, the trade between, and the modifications made as this action unfolds." This analytical approach permits dinosaurian representations being afforded the status as equally semiotic constructions and scientific evidence that should be discussed in continual dialogue with each other or, alternatively, as simultaneously "actual, once-living, scientifically verified, reconstructed, and always reconstructable beings" (Noble 2016: 18). In other words, examples of dino-visibility in peer-reviewed scientific journals, museums, genre literature or *Doctor Who* should be read as historically-determined instances that assist in performing particular ideological work. This work can reaffirm, complicate and develop long-standing discursive framings that make the Mesozoic meaningful:

> The Mesozoic works in this manner, as it has precursors, relies on conditions that make it sharable and effective with others, and opens to new circumstances that will allow it to be reconstituted, at least partially. Yet, whether illocutionary or perlocutionary, these performances no longer appear to be performances, or even strategic actions, because they have become so accepted as to hide the history and particularity of the practices informing them [Noble 2016: 37–8].

From this perspective, *Doctor Who*'s generically overdetermined mediations of dinosaurs are readable not as deliberate falsifications but as simultaneously appropriating and reworking long-established ideologies. For example,

> the articulation of *Tyrannosaurus* as ultimate aggressive killer, hunter, and carnivore had its beginnings between 1902 and 1917, through the rigorous collecting work of Barnum Brown, the engagement with several partial skeletons, and more pointedly though the scientific imagining and fossil interpretations of Henry Fairfield Osborn during his tenure at New York's American Museum of Natural History (AMNH) [Noble 2016: 71].

Whether denotative or otherwise, the depiction of tyrannosaurus-esque creatures in *Doctor Who* running from "The Silurians" to "Dinosaurs on a Spaceship" each appropriate this discourse which positions the lizard kings as "the ultimate foes to humanity" (Noble 2016: 101). It can therefore be argued that as "representations of animals are necessary for particular kinds of idea to be expressible" (Mills 2017: 6), encountering and overcoming these top predators from "deep time" (re)affirms the assumed "natural" superiority of typically white, scientifically educated, aged and aristocratically coded masculinity over an unruly, intrusive and savage natural world.

That the Doctor has, until their thirteenth incarnation, been constructed "as differing kinds of English gentleman" (Mills 2017: 46) reinforces this ideology. Alternatively, the narratively melodramatic codings of dinosaurs in new *Who* as innocent casualties of wider assailants works to maintain the other side of "the singular conception of the dinosaur as both pathetic, tragic victim and ruthless killer predators" (Sanz 2002: 35). Read from this perspective, new *Who* constructs dinosaurs "in terms of various lacks, and it is those lacks which are seen to justify particular kinds of treatment" (Mills 2017: 21)—namely the creatures' lack of intelligence equating to a failure in being able to adequately protect themselves, and so this responsibility must fall to white, male human characters. Within either version of dinosaur mediation, white masculinity's "natural" position as "superior" is frequently reasserted via the sympathetic characters' ability to rescue and avenge the mistreatment of prehistoric giants.

Of course, individual *Doctor Who* stories complicate these narratively-focused arguments through the lead character's alien status and "explicit liberalism" (McKee 2004: 203): the Third Doctor opposes the destruction of the Silurians and their creature, while also, in "Invasion of the Dinosaurs," wanting the dinosaurs transferred into the present by Operation Golden Age restored to their temporal home; the Fourth Doctor lets the Skarasen return to Loch Ness; the Eleventh Doctor ensures the safety of the Silurian ark's Mesozoic bounty in the face of hostility from human-coded characters like Solomon and Riddell; the T-rex's death motivates the Twelfth Doctor to investigate further the Half-Faced Man and unearth their sinister organ harvesting operation. It is only the Fifth Doctor who intentionally kills a dinosaurian character—and this comes with a wider expression of regret by the character at the point of overall narrative resolution. Rather than representing dinosaurs as assailants that must be ultimately vanquished to reaffirm white masculinity's dominance, *Doctor Who* instead asserts the importance of biodiversity and a caring scientific outlook. Although white masculinity as represented by the Doctor remains dominant in these instances—dinosaurs have so far been absent from the storylines of the female Thirteenth Doctor—it is not coded as deliberately forceful or aggressive within the context of this series.

However, the arguments developed throughout this essay suggest that refinements of Noble's (2016) position are required as greater sensitivity towards how specific contextual elements, such as representational medium and institutional milieu, highlight additional filters through which dinosaurian representations become structured. This is a perspective that Noble (2016: 40) demonstrates some sensitivity towards by arguing that invocations of the Mesozoic are "both historical (diachronically emergent and mobile) and heterogeneous (synchronically distributed and adjustable), and consequently a persistent, modern technique for generating normative yet revisable *natures*." However, what his point overlooks is how specific representations of dinosaurs and the Mesozoic are situated within wide-reaching-yet-fluid ideological structures *and* historically determined characteristics of the medium through which the dinosaurian representation is produced. As Glen Creeber (2013: 2) argues, "any medium is always in a state of potential change and development, influenced by technological innovation, economic, institutional and political pressures and so on." The discussion in this essay has demonstrated this point not only through noting how meanings assigned to dinosaurs have shifted to utilize these primarily for pre-publicity purposes, but also through changes in the level at which melodramatic associations become readable. Mills (2017: 6; original emphasis) offers support for the point I am developing here by arguing, albeit in an irreverent manner, that trying "to unpick what 'dogs' might mean is flawed; we can only make sense of what *this* dog representation means at *this* time given *these* contexts." Building upon this, I am arguing that we should examine not only what particular groups of dinosaurian representation tells us about the meaning of this category at a particular point in time, but also what that depiction tells us about the medium and organization that has produced the mediation. By offering such suggestions, more nuanced and context-sensitive accounts of how and why dinosaurs are made to mean in specific cases can be obtained.

Conclusions

This essay has argued that both denotative and connotative dinosaurian characters in *Doctor Who* have consistently been constructed as overdetermined by drawing upon combinations of narrative and aesthetic codes which position the creatures as simultaneously objects for scientific observation, dramatic spectacle, monstrosity, and melodramatic sentiment. The emphasis placed on how, when or whether these generic elements are accentuated has demonstrated changes over time such as shifting dinosaurs' visibility to the foreground for pre-publicity purposes within a competitive television marketplace or new *Who*'s attempts to downplay melodramatic readings to assist and secure the program's "quality" status. However, the continuities observable in *Doctor Who*'s dinosaurian constructions suggest that the series' approach to these creatures cannot be solely reduced to the "species-specimen" complex suggested by Noble (2016). Instead, a wider range of genre codings are identifiable. Additionally, the essay has argued that *Doctor Who*'s overdetermined mediations of dinosaurs should not be read as examples of scientific inaccuracies or a failure on behalf of the BBC to satisfy its public service responsibilities concerning informal learning in dramas engaging with (natural) history. In contrast, the discussion has argued that *Doctor Who*'s dinosaur mediations have simultaneously reaffirmed and challenged deep-rooted intertwinings of scientific and cultural ideologies concerning the Mesozoic by deploying melodramatic codings. These result in dinosaurs being positioned as creatures that should be respected, admired and looked after rather than functioning purely as ideological supports for the unquestioned dominance of characters coded as white, male, scientifically educated, aged and aristocratic.

The analysis in this essay has however been limited to televised *Doctor Who* stories featuring "giant lizard" creatures. Some consequences of this focus has been that analysis of Mesozoic-coded humanoid characters such as the Silurians and Sea Devils have been omitted as well as how dinosaurian characters have also featured in official media spin-offs such as the TV series *Torchwood* (2006–11), the novels *Blood Heat* (Mortimore 1993), *Scales of Injustice* (Russell 1996) and *The Silurian Gift* (Tucker 2013) and the Big Finish-produced audio dramas "Bloodtide" (2001) and "The Silurian Candidate" (2017). To expand upon the above call for adopting a medium specific understanding of the Mesozoic in popular culture, future discussion could use the representations offered by these characters and texts as points of comparison to examine the continuities and differences between "giant lizards" in televised *Doctor Who* and other media (paratexts). Undertaking this task would assist in considering whether dinosaurs within *Doctor Who* remain overdetermined as well as the ideological meanings of the Mesozoic which these either sustain or challenge.

The essay's arguments are significant as they ultimately indicate the difficulties of analyzing "science" as a discrete category through a popular long-running drama series such as *Doctor Who* and its status as a television program. Robin Nelson (2007: 21) has argued that "Mixes of different genres are not new" within the creation of television shows and Jason Mittell (2004: 155) has expanded this point by highlighting that "Fusion can occur at a variety of levels: individual episodes […] specific programs […] or emergent genres (the late 1980s rise of 'dramedies' fusing dramas and sitcoms)." In contrast, this argument has highlighted individual sequences in *Doctor Who* where objects of scientific interest such as dinosaurs are overdetermined in terms of genre through being open to multiple genre classifications simultaneously. Following Noble (2016), this should not lead to such representations being dismissed as "unscientific," however. Instead, I would argue that the example of analyzing the history of dinosaurian constructions within *Doctor Who* demonstrates that any discussions of "science," as a televisual category, needs to demonstrate greater sensitivity to the aesthetic characteristics, narrative codings and medium characteristics that structure its mediation as well as the wider ideologies which such mediations draw from.

NOTES

1. The Silurians' dinosaur character is never described as belonging to a specific species within the transmitted story. However, given the creature's bipedal nature, its coding as a threat to humanoid characters, and how "no scientific name is as well known as *Tyrannosaurus rex* ... and the animal that bears that name is easily the most popular and best-known dinosaur with the general public" (Hone 2016: 23), it is arguable it can be read as a T-rex.

REFERENCES

BBC Studios (2019, March) *BBC Annual Plan 2019–20*. downloads.bbc.co.uk/aboutthebbc/reports/annualplan/annualplan_2019-20.pdf (accessed 7 October 2019).
Brunsdon, C. (1997) *Screen Tastes: Soap Opera to Satellite Dishes*. London: BFI.
Cardwell, S. (2002) *Adaptation Revisited: Television and the Classic Novel*. Manchester: Manchester University Press.
Carroll, N. (1990) *The Philosophy of Horror, or Paradoxes of the Heart*. London: Routledge.
Caughie, J. (2014) What do actors do when they act? In: Bignell, J., and Lacey, S. (eds.) *British Television Drama: Past, Present, and Future*, 2nd edition. Basingstoke: Palgrave MacMillan, pp. 143–158.
Chapman, J. (2013) *Inside the TARDIS: The Worlds of* Doctor Who, 2nd edition. London: I.B. Tauris.
Creeber, G. (2013) *Small Screen Aesthetics: From TV to the Internet*. London: BFI.
DCMS (2016, December) *Royal Charter for the Continuance of the British Broadcasting Corporation*. www.gov.uk/government/publications/bbc-charter-and-framework-agreement (accessed 7 October 2019).
Dunleavy, T. (2009) *Television Drama: Form, Agency, Innovation*. Basingstoke: Palgrave Macmillan.
Fish, S. (1980) *Is There a Text in This Class? The Authority of Interpretive Communities*. Cambridge: Harvard University Press.
Foucault, M. ([1989] 2003) *The Birth of the Clinic: An Archaeology of Medical Perception*. London: Routledge.
Frank M. (2013) At the intersections of mode, genre, and media: A dossier of essays on melodrama. *Criticism* 55(4): 535–545.
Hendy, D. (2013) *Public Service Broadcasting (Key Concerns in Media Studies)*. Basingstoke: Palgrave.
Hills, M. (2010) *Triumph of a Time Lord: Regenerating* Doctor Who *in the Twenty-First Century*. London: I.B. Tauris.
Hone, D. (2016) *The Tyrannosaur Chronicles: The Biology of the Tyrant Dinosaurs*. London: Bloomsbury Sigma.
Jermyn, D. (2007) Body matters: Realism, spectacle and the corpse in *CSI*. In: Allen, M. (ed.) *Reading* CSI: *Crime TV Under the Microscope*. London: I.B. Tauris, pp. 79–89.
Johnson, C. (2005) *Telefantasy*. London: BFI.
Lacey, S. (2016) Just plain "odd": Some thoughts on performance styles in *Twin Peaks*. *Cinema Journal* 55(3): 126–131.
Lury, K. (2005) *Interpreting Television*. London: Hodder Arnold.
McKee, A. (2004) Is *Doctor Who* political? *European Journal of Cultural Studies* 7(2): 201–217.
Mills, B. (2017) *Animals on Television: The Cultural Making of the Non-Human*. Basingstoke: Palgrave MacMillan.
Mitchell, W.J.T. (1998) *The Last Dinosaur Book: The Life and Times of a Cultural Icon*. Chicago: University of Chicago Press.

Mittell, J. (2004) *Genre and Television: From Cop Shows to Cartoons in American Culture.* London: Routledge.
Mortimore, J. (1993) *Doctor Who: Blood Heat.* London: Virgin Publishing Ltd.
Nelson, R. (2007) *State of Play: Contemporary "High-End" TV Drama.* Manchester: Manchester University Press.
Newman, K. (2005) *BFI TV Classics: Doctor Who.* London: BFI.
Noble, B. (2016) *Articulating Dinosaurs: A Political Anthropology.* Toronto: University of Toronto Press.
Pearson, R. (1992) *Eloquent Gestures: The Transformation of Performance Style in the Griffith Biograph Films.* Berkeley and Los Angeles: University of California Press.
Petley, J. (2006) Public service broadcasting in the UK. In: Gomery, D., and Hockley, L. (eds.) *Television Industries.* London: BFI, pp. 42–45.
Russell, G. ([1996] 2014) *Doctor Who: Scales of Injustice.* Croydon: BBC Books.
Sanz, J.L. (2002) *Staring T.Rex!: Dinosaur Mythology and Popular Culture.* Bloomington and Indianapolis: Indiana University Press.
Sleight, G. (2012) *The Doctor's Monsters: Meanings of the Monstrous in Doctor Who.* London: I.B. Tauris.
Telotte, J.P. (2001) *Science Fiction Film.* Cambridge: Cambridge University Press.
Tucker, M. (2013) *Doctor Who: The Silurian Gift.* Croydon: BBC Books.
Turnbull, S. (2007) The hook and the look: *CSI* and the aesthetics of the television crime series. In: Allen, M. (ed.) *Reading* CSI: *Crime TV Under the Microscope.* London: I.B. Tauris, pp. 15–32.
Vint, S. (2010) *Animal Alterity: Science Fiction and the Question of the Animal.* Liverpool: Liverpool University Press.
Wheatley, H. (2016) *Spectacular Television: Exploring Televisual Pleasure.* London: I.B. Tauris.

The Use and Abuse of Scientific Writing in *Doctor Who*'s Epistolary Paratexts

Tonguç İbrahim Sezen

Introduction

In 2018, issue 117 of the popular science and technology magazine *How It Works* featured an article on the science and technology of *Doctor Who*. Describing various scientific concepts which inspired the series such as general relativity, nuclear fusion and cell regeneration with easy-to-understand texts and diagrams, the article argued that *Doctor Who* has been carefully interweaving science and science fiction for years "to create an entertaining combination of fantasy and reality, supported by somewhat accurate depictions of various scientific principles" (Dutfield 2018: 23). With additional behind-the-scenes information covering the development of the series and comments on the differences between the fictional and actual states of various technologies, the article was a typical extra-diegetic paratext like making-of or the art-of books aiming to enhance the franchise's cultural status (Hills 2019), in this case by relating *Doctor Who* to current scientific and technological advances. One of the key features of the article was the inclusion of technical cutaway diagrams of popular cybernetic *Doctor Who* villains the Daleks and the Cybermen as well as a cutaway of the first TARDIS control console. These were drawn in a style not unlike the diagram of a World War I era German A7V tank appearing in the same issue, with technical descriptions of various components such as "a network of hydraulic pipes control the wires and cables that dictate the movement of the emotionless Cybermen," or "located just behind the eye lens, this receptor enables a Dalek to detect which direction a sound is coming from," and "the [Artron] mainframe connects all the computer networks and systems aboard the TARDIS and acts as its interface for the Doctor" (Dutfield 2018: 26–30). These diagrams were contributing to the verisimilitude of *Doctor Who* by giving the impression that these alien technologies really existed and were functioning in a reasonable fashion based on plausible scientific principles. In this regard, they could be considered as intra-diegetic elements (Ryan 2017), meaning they gave the impression that they could exist within the story world of *Doctor Who*, even though they were not part of its narrative. These diagrams were functioning as what we may call "epistolary paratexts," only one of many from the long history of *Doctor Who*. In this regard this essay will first define the concept and later discuss the use and in some cases abuse of epistolary paratexts in presenting the fictional world of *Doctor Who* as a separate, alternative reality with a functioning and plausible infrastructure.

Defining Epistolary Paratexts

Epistolary paratexts can be defined as ancillary reference material accompanying narrative urtexts and/or transmedia corpuses, and claiming to originate from the same imaginary worlds these texts are set in. As in Hynes' interpretation of Genette's original conception of literary paratexts within the context of imaginary worlds (Hynes 2018), epistolary paratexts too form a threshold around narrative texts to mediate their imaginary worlds to the audience. But while Hynes argues that paratexts do not form part of the diegetic world itself, epistolary paratexts are intra-diegetic. Like epistolary novels or fictional films of the found footage genre, they claim to be authentic documents originating from a fictional world. But unlike these genres, epistolary paratexts are not specifically organized to form a fictional narrative. Any narrative they form is derived from their mimicry of factual narration. Usually in the form of notebooks, scientific reports, engineering manuals, technical drawings, and similar reference documents, epistolary paratexts aim to systematically organize the data on the inner workings of a fictional world originally communicated through narrative texts. While they tend to retell or reorganize information provided by narrative texts, they may also introduce additional data to establish a coherence. Masquerading as if they were written by characters inhabiting those fictional worlds, they may also provide personal or institutional perspectives.

Most contemporary epistolary paratexts are products of the evolution of cross-media and transmedia practices blurring the line between reality and fiction. Christy Dena (2009) describes this new type of aesthetics as verisimilitude. It allows real world audiences to interact with fictional worlds through constructs existing on both worlds (Blumenthal 2016). In this regard, Henry

Jenkins (2009) argues that many parallels can be drawn between the tradition of epistolary novels and transmedia texts since they both invite their audiences to construct a fictional reality from fragments. Epistolary paratexts can act as one of these fragments or take a referential role trying to give an overview of various aspects of these fragmented realities.

Precursors of epistolary paratexts can be traced back to late-nineteenth century in New Romance adventure novels. According to Michael Saler (2012), the use of footnotes, glossaries, appendices, maps, and tables claiming to be authentic artifacts in novels such as H. Rider Haggard's *King Solomon's Mines* (1885) had contributed to the transformation of literary imaginary worlds into cohesively structured, empirically detailed, and logically based "virtual realities." Following J.R.R. Tolkien's influential use of maps of the Middle-earth in *The Lord of the Rings* (1954–55), inclusion of maps depicting fictional geographies has become a common practice in fantasy literature, which Sally Bushell (2016) argues created an almost symbiotic relationship between the visual and verbal contents. In the late 1950s the *Eagle Magazine* popularized technical cutaway diagrams of fictional space vehicles by publishing diagrams based on its science fiction comic series *Dan Dare* (Sillars 1995). Throughout the 1970s reference-oriented fans codified and elaborated *Star Trek*'s (1966–69) fictional world through blueprints, concordances and technical manuals that led to the release of the franchise's first official reference books exploring the universe in the show's backdrop as twenty-third century Starfleet publications (Rehak 2016). By the early 2000s, Jane McGonigal (2003) coined the term "immersive aesthetics" to describe the conventions of the emerging alternate reality game genre, like the creation of complex networks of online diegetic content for players to explore. Immersive aesthetics would later be adopted by various movie and TV franchises to establish diegetic online presences for marketing purposes, some of which would resemble technical guidelines and training documents (Tavares 2015). Finally, contemporary multimodal novels would try to adapt these techniques back into the print format by imitating catalogs or scrapbooks (Gibbons 2017).

Currently, digital and printed epistolary paratexts are common tie-in products of many major media franchises. Targeting mainly reference (Rehak 2016) and design oriented (Rehak 2018) fans, such publications combine the object-oriented materiality of extra-diegetic reference sources, i.e., making-of and art-of books, with the make-believe of fictional narratives. Using factual, technical, and scientific language, they contribute to the perception of fictional worlds as functional and plausible realities. Since 1964, *Doctor Who* too has been accompanied by such texts exploring and expanding its fictional world through the lens of encyclopedic and technical writing.

World-builder's Chronicles

The history of epistolary paratexts of the *Doctor Who* franchise can be traced back to the second serialized *Doctor Who* story, "The Daleks" (1963–64) written by Terry Nation, which was the first *Doctor Who* serial to have a futuristic setting. The titular Daleks were merciless cyborg aliens reigning over the post-apocalyptic planet of Skaro from within their city and dependent on static electricity for movement, and oppressing a group of pacifist humanoids called the Thals. Thanks to the unusual design and fearful character of the Daleks, the serial became an overnight success and caused great excitement among younger audiences demonstrated by fan mail expressing children's desire to be immersed in new Dalek stories in future *Doctor Who* episodes (Bignell and O'Day 2004). Later dubbed "Dalekmania," this excitement and demand first surprised and later encouraged the BBC to produce more Dalek centered episodes and even two theatrical movie adaptations, even though the Dalek species appeared to have been killed off by the end of the original serial (Chapman 2014). Soon multiple tie-in products and merchandise including toys, games, comics, and books followed. Due to his copyright ownership of the Daleks, Terry Nation not only commercially benefited from this popularity, but was also involved in most of the Dalek related production and even was able to sign deals for books on Daleks without the direct involvement of the BBC (Turner 2011). Between 1964 and 1966, Nation co-authored a series of books published by the Souvenir Press featuring the Daleks and his other creations from various *Doctor Who* episodes such as the Mechonoids and agent Sara Kingdom, but not the Doctor. These titles were *The Dalek Book* (Whitaker and Nation 1964), *The Dalek World* (Whitaker and Nation 1965), *The Dalek Pocketbook and Space Travellers Guide* (Nation 1965), and *The Dalek Outer Space Book* (Nation and Ashton 1966). Following the format of comic book annuals of the era, such as *Dan Dare's Space Annual 1963* (No Author 1962), they were essentially collections of comic strips, short stories and puzzles focusing on Daleks' conflicts with future human civilizations and other alien cultures.

With Nation's involvement in both Souvenir Press' and the BBC's creative processes, the Dalek books were a test bed for world building experiments on *Doctor Who*. Their authors were trying to create a solid and somewhat separate Dalek mythology by deliberately introducing and describing various elements of their universe, which as Turner (2011) notes hopefully would create the basis of and the interest in other Dalek related products. Despite and maybe because of their outlandish content, these books were marketed as chapters of the authentic future-histories of the Dalek civilization supposedly based on the mysterious "Dalek Chronicles" Nation had discovered and translated. While most of these adaptations were in the form of comic books and short stories, the books also contained multiple scholarly apparatus such as technical illustrations, maps, dictionaries, timelines and articles on the culture, biology, and technology of the Daleks. Albeit fragmented across multiple books, these were the first epistolary paratexts of the *Doctor Who* franchise using factual formats to deliver fictional content and claiming authenticity.

Epistolary paratexts are specifically used to organize and deliver assorted information about a fictional or secondary world in a coherent and consistent manner (Wolf 2012), but usually this information exists in preceding narrative texts. In the case of Souvenir Press' Dalek books, these were used to introduce the secondary world. Mark J.P. Wolf (2012) argues that each secondary world requires three basic elements to exist: a space, a timeframe, and characters to inhabit them. Based on the complexity of the secondary world and its level of divergence from our prime world, other elements including nature (and technology based on materiality), culture, language, mythology, and philosophy may be described as well. Plausibility and more importantly consistency in describing and using these elements bears great importance. In this regard, to help readers to imagine a functional and continuous *Doctor Who* universe, epistolary paratexts in Dalek books provided data on five of the eight of these world building elements: space, time, nature, culture and language. In the first three areas they were clearly trying to establish a scientifically and technologically plausible world image.

As the main setting of many Dalek stories, the planet Skaro was covered throughout Dalek books in three different documents: a more or less realistic geographical map of the planet, a somewhat fantastical cutaway diagram of the planet's strata mainly focusing on underground life forms, and finally an article on the city of Daleks and the dangers it poses for non–Dalek visitors. According to Bushell (2016) maps accompanying fictional narratives help readers to interpret narratives but also raise questions. This is also true for the geographical documents in Dalek books. The planet's map reveals names and positions of the continents populated by Daleks and the Thals and shows the damage nuclear war has done, thus provides with the help of scaling and grid references a plausible global context to the original Dalek serial. It also raises questions and opens possibilities for future stories by introducing unexplored areas, such as the misty Forbidden Islands. Reprinted multiple times (Nation 1965, Mann et al. 2017), this map was one of the first epistolary paratexts using a factual format to establish a plausible fictional world image. While fulfilling the same narrative functions, the strata cutaway in *The Dalek Outer Space Book* provided a less plausible more fantastical image. The planet's crust was shown to have multiple layers with unique characteristics and flora and fauna: an extremely hot outer layer was populated with animals covered with furs made of asbestos, another layer of living sponge was populated by sponge people, and finally the planet's core was frozen and was expected to form ice-volcanoes throwing up super cold ice, an idea which would later be used in "Planet of the Daleks" (1973) for the Dalek base planet Spiridon. Despite the differences regarding their plausibility, the shared motive among all these geographical epistolary paratexts was the uncanniness of the Dalek home world with its Ocean of Death, Zone of Eternal Dark, and Canyons of Terror.

The second core world-building element the Dalek books' epistolary paratexts expanded upon was time, not the timeframe of Dalek stories but the history of the Dalek civilization. A timetable in the *Dalek Outer Space Book* summarized in one page the evolution of life, the emergence of technologically advanced societies, the period of global war, and the rise of the Daleks on the planet Skaro. Mirroring the evolution of life on earth from single celled organisms to giant reptilian animals and finally to humanoids, this timetable was almost suggesting convergent evolution on a galactic scale. In this regard, the war-torn latter part of this history that led to the creation of Daleks could be seen as a warning to the readers as well. Moreover, the timeline was also vaguely foreshadowing future *Doctor Who* episodes on the genesis of the Daleks without giving any details. Thus, it was both creating a plausible historical background for the Daleks as a species and providing narrative cues for future stories.

Another source on Dalek history was an article in *Dalek World* listing random facts about the aliens. These "facts" were trying to explain historical events such as the construction of the Great Pyramid of Giza through Dalek incursions to the earth. These were among the first examples of pseudo-historical epistolary paratexts in *Doctor Who* literature, which were mixing history with science-fiction elements and proposing alternate fictional explanations to historical records and archaeological findings. As Andrew O'Day (2017) points out, pseudo-historical narratives would become a staple of *Doctor Who* after the 1965 serial "The Time Meddler" (1965), and the way historical facts and theories are presented would be more and more affected by the requirements of the stories. As it will be seen below, the pseudo-historical approach would become more prominent in future epistolary paratexts as well and this would cause some unexpected conflicts.

The main area the Dalek books covered with the highest number of epistolary paratexts was the nature and technology of the Daleks. These texts were introducing minor and major "novums." Proposed by Darko Suvin (1979), the term novum refers to describe scientifically plausible innovations in science fiction narratives that differentiate the world of fiction from its readers' reality. To describe these novums, Dalek books' epistolary paratexts used cutaway diagrams showing the internal parts of vehicles or devices and illustrating their relationship to the whole (Alred et al. 2015) or technical definitions explaining unfamiliar concepts (Balzotti 2018) with various levels of detail. Some of these novums were vaguely described and were more fantastical than scientific in nature. For example, the doomsday weapon called Anti-Solarium, introduced in *The Dalek Book* feature "The Dalek War Machines," looked like a radar dish and supposedly could inject a system of refrigeration into the rays of a star and then use it to freeze planets. Similarly, the cutaway diagrams for the Dalek spaceship, or the Skaro Saucer, and the Dalek submarine, or the Dalesub, were confusingly claustrophobic, disconnected and out of scale compared to their given capabilities, much like the diagram of Skaro's strata. Yet there were also epistolary paratexts describing more plausible novums. The previously mentioned list of random facts about the Daleks for example gave multiple details about their speed, weight, energy production, color perception, and cognition. While some of these facts such as Daleks not being able to see the color red conflicted with their portrayal in other media, these were consistent and plausible within the range of the Dalek books.

The most prominent and detailed cutaway diagram in Dalek books though was the Anatomy of a Dalek, which showed the inside of a Dalek (something only tantalizingly hinted at on screen in the TV series) for the first time among all *Doctor Who* media. It introduced many elements which would be kept in future iterations such as the above mentioned *How It Works* version: a technical design matching the outer shell with repetitions and symmetry, internal features enabling the perceived functionality of the Daleks, and an internal segmentation matching the fictional description of Daleks' cybernetic nature. The diagram provided the basics of the system topology of a Dalek, a model of how its internal components are connected and how they work together, which according to Kieras and Bovair (1984) is the key component for readers to understand a device's functionality even if it originates from a science fiction setting. As the only technical epistolary paratext in Dalek Books directly based on a design from the TV show, the Anatomy of a Dalek was enhancing the on-screen design with functional, consistent and plausible novums and thus contributing to the verisimilitude of the world of *Doctor Who*.

Dalek books' purpose to build up a functional and breathing fictional world was probably most evident in *The Dalek Pocketbook and Space Travellers Guide*, which claimed to be an actual guide on Daleks and their civilization. While most of its epistolary paratextual content consisted of reprinted material from earlier Souvenir Press publications, it also included chapters on by then state of the art space flight technologies and space exploration proposals by contemporary scientists. While these were separated as chapters, their juxtaposition in the same book contributed to their claim of authenticity. Souvenir Press' Dalek books were amongst rare examples of epistolary paratexts' use as an active component of world building at an early phase of a franchise's growth. They were introducing elements and trying to present them as plausible natural phenomena or scientific or technological novums to build audience interest, which would be the prime goal of most future *Doctor Who* epistolary paratexts as well.

An Archaeological Approach

Following Nation's world-building experiment, official *Doctor Who* periodicals and specials would rarely publish epistolary paratextual chapters. Instead, they would archive the on-screen continuity of *Doctor Who* through behind the scenes articles, photo galleries, and episode summaries. Themed specials (Nation 1978; Nation 1979) would feature technical diagrams and descriptive articles, but they still would refrain from claiming authenticity. As an exception, the *Doctor Who Yearbook 1993* (No Author 1992) would feature a collection of "Monster Files" and "Anatomy of..." cutaway drawings of various alien races from different eras of the show. This unique yearbook in many ways reflected a shift in *Doctor Who* publications that had started in the late 1970s and early 1980s as an answer to changing viewing conditions.

According to Matt Hills (2013) the original run of *Doctor Who* could be divided into two periods in terms of audience access to past episodes: the scarcity period throughout the 1960s and 1970s, which can be characterized by the rarity of reruns and the technological impossibility of recording or commercially releasing past episodes, and the home video period of surviving episodes starting in the 1980s which led to the formation of socially-organized fan communities regularly cataloguing and debating the show. In discussing the effects of this shift on the creative processes behind *Doctor Who*, Mark Bould (2012) argues that the serials written during the scarcity period lacked overarching story arcs and continuous memories from one serial to the next due to their unrecoverable nature and were filled with details introduced on an ad hoc basis. Whereas, throughout the home video period *Doctor Who* would become a series more and more relying on its past, despite its multiple inconsistencies inherited from the scarcity period. In this regard Bould acknowledges the emergence of a "desire to give *Doctor Who*'s story-world the kind of coherence typical of more contemporary franchises" (Bould 2012: 153). Epistolary paratexts in the form of reference books would be used to answer this need.

The earliest *Doctor Who* themed reference books can be traced back to the mid–1970s. Covering various aliens from the *Doctor Who* universe with in-universe descriptions and a mixture of paintings and still images from the show, *The Doctor Who Monster Book* (Dicks 1975) may be considered one of the first examples of such publications. As Jonathan Appleton (2014) points out, the main feature of the book was its coverage of aliens dating back to the earliest days of the scarcity period, which for most fans of the time was their only introduction to this content. Throughout the 1980s and 1990s *Doctor Who* reference books became more diverse and complex. *The Doctor Who Technical Manual* (Harris 1983) was a collection of scaled plans, drawings and cutaways of various mechanical aliens, vehicles and devices. Written in a design-oriented in-universe style, the book juxtaposed multiple iterations of aliens from different serials as their different types and thus acknowledged the design history of the series. *The Doctor Who Illustrated A-Z* (Standing 1985) was written as a short, illustrated encyclopedia with entries on various aspects of the *Doctor Who* universe with internal references connecting related content. *Doctor Who: Cybermen* (Banks 1990) mixed epistolary paratextual writing with extradiegetic paratextual writing and provided an in-depth coverage of the evolution of the Cybermen both in fiction and behind the screens in a way not unlike Souvenir Press' Dalek books, but by relying on the show's history instead of introducing new content.

As epistolary paratexts, these books aimed to organize a fragmented fictional universe. They were covering and organizing content from the vast narrative of *Doctor Who*, some dating back to the earliest days of the scarcity period, for home video era fans to study. These books were not trying to actively build the world of *Doctor Who*, but by reintroducing elements previously considered unrecoverable and relating them with recent elements they were contributing to the build-up of a continuous fictional universe. In the foreword of the online version of *Doctor Who: The Universal Databank* (Lofficier 1992), the author Jean-Marc Lofficier describes the goal of such reference books as creating a "retroactive continuity" by analyzing past *Doctor Who* episodes retroactively and organizing them into a coherent continuity. Pointing out multiple difficulties inherent in *Doctor Who* as a TV show, including its vastness and many inconsistencies between serials, Lofficier argues that this retroactive continuity cannot be built only based on information given in the show but also requires interpretation and creativity. Lofficier (n.d.) proposes a set of rules to be used in creating retroactive continuities, which can be summarized as follows:

    1. Identifying limitations, primary and secondary sources.
    2. Privileging historical accuracy and up-to-date scientific information.

3. Providing relevant bits of scientific or historical data to establish context.
4. If fiction conflicts with scientific information, trying to reconcile fiction with science as much as possible.
5. If there are conflicts among fictional texts, trying to reconcile these, usually by privileging the simplest answer.
6. If there are missing factual details estimating these by reviewing existing data.
7. Introducing as little as possible new fictional content to solve continuity problems or plot holes.

Lofficier's rules resemble both fans' forensic mode of engagement with contemporary TV shows, which leads to the formation of fan-wikis (Mittell 2007), and the completion of necessary gaps in various world infrastructures of urtexts, often required to author licensed reference books (Wolf 2012). Lofficier himself compares his and his contemporaries' work on *Doctor Who* to archaeology and underlines how different authors' interpretation of the same data may lead to different conclusions. In this regard, Paul Cornell's (2007) account on how some of his conclusions on Daleks' history in *The Discontinuity Guide* (Cornell et al. 1995) were silently rejected by fans shows the hypothetical nature of these texts.

Focusing on licensed novels and growing fan production in the 1990s, Neil Perryman (2008) argues that despite its long history of licensed products, *Doctor Who* only truly started to become a transmedia franchise after the cancellation of the original series in 1989. This period also coincides with the growing publication of reference books with in-universe content. Before the emergence of wikis and other folksonomies, reference books like these were providing fans plausible and coherent theories to discuss and build their own vision upon by cataloguing and organizing the world infrastructure of *Doctor Who*. Following the relaunch of *Doctor Who* in 2005, the number of reference books would continue to increase. Aiming at different age groups, these books were both covering the ongoing series and introducing the long history of the franchise to new audiences. Multiple reference books mixing epistolary paratextual and extradiegetic paratextual writing connected various elements from both *Doctor Who* series' by summarizing their on-screen and production history. Other books, such as *Doctor Who: The Visual Dictionary* (Loborik *et. al* 2015) followed Lofficier's encyclopedic approach, but maybe because of the competition of fan-wikis, with a special emphasis on visuals. All in all, since the late 1970s, with their gradually evolving methodological and almost scientific approach, in-universe reference books were treating the secondary world of *Doctor Who* as an authentic entity. What they mostly lacked was a claim on authenticity of their own.

Authentic Documents from Another Universe

Throughout the original run of the series and the following interregnum period, epistolary paratexts openly claiming authenticity were a rarity. In the mid–1980s the US based game company FASA published a series of sourcebooks to accompany their *Doctor Who* themed roleplaying game, some of which were written as a series of intelligence documents on various alien threats including the Daleks, Cybermen, and the Master. These documents were both collecting and organizing data from the TV show, but also adding new practical content to provide players material support in exploring and inhabiting the game world. *The Dalek Problem: A Symposium* (No Author 1985), a document written supposedly by a Professor Qualenawitvanastech from the Multihistorical Research department, for example contained well-organized, in-depth chapters on Dalek physiology, technology, history, society and military, and ended with a conclusion and recommendations. Besides imitating the structure of an academic document, *The Dalek Problem* was purposefully opening possibilities for players to explore and react on. In other words, FASA epistolary paratexts were designed to be used to create an iteration of the on-screen world of *Doctor Who*, which would come into life through players' actions.

Another rare example was *The Dalek Survival Guide* (No Author 2002), published near the end of the interregnum period between the two series. At first glance the book seemed to be an updated version of the *Dalek Pocketbook*, but despite collecting a wide range of data on the Daleks and using multiple diagrams to visualize them, the book was in fact a humorous, tongue-in-cheek spoof of surviving an encounter with the cybernetic aliens. Its "tips" for survival included looking for exit signs, not standing in front of a Dalek, and riding one in a somewhat erotic fashion to avoid extermination; all jokes which could only work in an epistolary paratextual format. What *The Dalek Survival Guide* introduced to *Doctor Who* epistolary paratextual writing was its tone; dry, straightforward, scientific but at the same time humorous and sarcastic, a voice which could turn epistolary paratexts into more appealing texts than the scientific documents they were mimicking.

Following the 2005 relaunch of the series, two main types of epistolary paratexts emerged. The first one was the serious, official documentation type that sometimes mimicked Gallifreyan books and manuals. One of them, *TARDIS Type 40 Instruction Manual* (Atkinson and Tucker 2018) was a collection of TARDIS related bits of information collected from both series. Organized categorically and in a coherent fashion, the book imitated an actual vehicle manual with chapters on subjects like engines, circuits, and common malfunctions. The text was written in an official, dry manner, and was accompanied by multiple diagrams, images and cutaways as well as summaries of TARDIS centered episodes as case files. Its thematic narrowness was one of the key factors in creating a balanced and fulfilling content, much like the previous *Dalek Pocketbook* or FASA sourcebook supplements. All in all, the book was an evolved version of previous epistolary paratextual documents organizing novums in functional, consistent and plausible ways to create a believable secondary world infrastructure for *Doctor Who*.

The second common type of post–2005 epistolary paratexts were the ones using "scrapbook aesthetics" or integrated personal commentaries to give the Doctor a voice while presenting data from the show. One of these books, *The Dangerous Book of Monsters* (Richards 2017), looked like the Doctor's personal scrapbook collecting images, drawings, documents, and handwritten notes on various aliens from the series. While its thematic organization resembled encyclopedic reference books, it was more akin to multimodal novels (Gibbons 2017), due to its use of the Doctor's handwritten notes for directly addressing the readers. These notes were not only giving data such as survival tips, but also representing the Doctor's personality and mannerisms, which within the context of *The Dangerous Book of Monsters* contributed to the claim of authenticity of the book.

But scrapbook aesthetics could also lead to unintended problems regarding the presentation of a plausible world and its connection to the prime world. These problems can best be seen in a comparison between *Doctor Who: The Whoniverse* (Mann and Richards 2016) and *Doctor Who: A History of Humankind: The Doctor's Official Guide* (No Author 2016), both covering historical events, including the Big Bang, emergence of life on earth, extinction of dinosaurs, and evolution of humans, by summarizing the pseudo-historical narratives of *Doctor Who*. While *The Whoniverse* presented these in a neutral manner from within the world of *Doctor Who*, *A History of Humankind*'s tone was more aggressive and mocking. Claiming to be a textbook from the fictional Coal Hill School Library, *A History of Humankind* was supposedly edited by the Doctor, who had replaced prime world facts and theories with pseudo-historical narratives and turned it into a scrapbook. Unlike juxtapositions of fact and fiction in previous epistolary paratexts, which managed to keep these separated but mutually supportive, *A History of Humankind* deliberately rejected prime world scientific knowledge. The Doctor's notes on the extinction of dinosaurs read for instance, "There were a couple of pages here giving all sorts of daft theories about why the dinosaurs became extinct—stuff about meteors and asteroids and comets. All complete rubbish." (No Author 2016: 12). While both books were recounting the same episodes, *A History of Humankind*'s approach breaks the threshold between the two worlds and openly privileges fiction over science. *The Whoniverse*'s approach on the other hand may appear distant by not directly addressing the readers but presents an alternate continuous and plausible fictional universe without appearing to be counterfactual.

Conclusion

Claiming to be authentic documents from another universe, epistolary paratexts have been part of *Doctor Who* since its infancy. Over the years, they contributed to the buildup and consumption of the franchise in multiple ways. Terry Nation's world building experiment had shown how apparently scientific documents could be used to create a fictional world. Their visual and textual format paved the way for more advanced and complex examples. While not openly claiming authenticity, by approaching the world of *Doctor Who* with an almost scientific attention, reference books helped fans to organize and discuss the consistency of a traditionally canonless franchise. Their contribution to the verisimilitude of the series came not from their creativeness but from the way they study and interpret its history, culture and nature. Contemporary epistolary paratexts build on these previous approaches and provide new ways to explore the world building elements of the series. They study, organize and fill in the gaps, and even comment on it, which in some cases leads to clashes between the worlds.

Epistolary paratexts do not exist by themselves. They require urtexts to explore and explain. Their contribution to a fictional world does not come from their recounting of it, but from organizing, thematically classifying, and providing the logic and science behind its working. Epistolary paratexts show what's inside of a prop even if the production team had never created it. They question and provide answers to bits of technobabble dialogue. They remember and relate the differences in the looks and actions of aliens. At the same time, they are also ephemeral content that may be declared false by every new story. They have great competition in the form of fan-wikis and fan-art. They are, in the end, interpretations of fictional worlds for audiences who want to take a step back and see the world behind the stories, wondering how it works.

References

Alred, G.J., Brusaw, C.T., and Oliu, W.E. (2015) *The Handbook of Technical Writing, Eleventh Edition.* Boston: Bedford/St. Martin's.
Appleton, J. (2016) Our favorite merchandise: The *Doctor Who* monster books. *The Doctor Who Companion.* thedoctorwhocompanion.com/2016/06/30/our-favourite-merchandise-the-doctor-who-monster-books/ (accessed 15 November 2019).
Atkinson, R., and Tucker, M. (2018) *TARDIS Type 40 Instruction Manual.* London: BBC Books.
Balzotti, J. (2018) *Technical Writing Essentials.* Provo: BYU Academic Publishing.
Banks, D. (1990) *Doctor Who: Cybermen.* London: W.H. Allen & Co.
Bignell, J., and O'Day, A. (2004) *Terry Nation.* Manchester: Manchester University Press.
Blumenthal, H. (2016) *Storyscape: A New Medium of Media.* Ph.D. dissertation, Georgia Institute of Technology.
Bould, M. (2012) *Doctor Who*: Adaptations and flows. In: Telotte, J.P., and Duchovnay, G. (eds.) *Science Fiction Film, Television, and Adaptation.* New York: Routledge, pp. 143–163.
Bushell, S. (2016) Paratext or imagetext? Interpreting the fictional map. *Word and Image* 32(2): 181–194. doi:10.1080/02666286.2016.1146513.
Chapman, J. (2014) Fifty years in the TARDIS: The historical moments of *Doctor Who. Critical Studies in Television: The International Journal of Television Studies* 9(1): 43–61. doi:10.7227/CST.9.1.4
Cornell, P. (2007) Canonicity in *Doctor Who.* Paul Cornell. www.paulcornell.com/2007/02/canonicity-in-doctor-who/ (accessed 10 November 2019).
Cornell, P., Day, M., and Topping, K. (1995) *The Discontinuity Guide.* London Virgin Books.
Dena, C. (2009) *Transmedia Practice: Theorising the Practice of Expressing a Fictional World across Distinct Media and Environments.* Ph.D. dissertation, University of Sydney.
Dicks, T. (1975) *The Doctor Who Monster Book.* London: Target Books.
Dutfield, S. (2018) The science and tech of *Doctor Who. How it Works* 117: 22–31.
Gibbons, A. (2017) Reading S across media: Transmedia storyworlds, multimodal fiction, and real readers. *Narrative* 25(3): 321–341. doi:10.1353/nar.2017.0017.
Harris, M. (1983) *The Doctor Who Technical Manual.* London Severn House.
Hills, M. (2013) Anniversary adventures in space and time: The changing faces of *Doctor Who*'s commemoration. In: Hills, M. (ed.) *New Dimensions of Doctor Who: Adventures in Space, Time and Television.* London: I.B. Tauris, pp. 216–234.
Hills, M. (2019) Transmedia paratexts: Informational, commercial, diegetic, and auratic circulation. In: Freeman, M., and Gambarato, R.R. (eds.) *The Routledge Companion to Transmedia Studies.* New York: Routledge, pp. 289–296.
Hynes, G. (2018) Geography and maps. In: Wolf, M.J.P. (ed.) *The Routledge Companion to Imaginary Worlds.* New York: Routledge, pp. 98–106.
Jenkins, H. (2009) Revenge of the origami unicorn: The remaining four principles of transmedia storytelling. *Confessions of an Aca-Fan.* henryjenkins.org/blog/2009/12/revenge_of_the_origami_unicorn.html (accessed 16 October 2019).
Kieras, D.E., and Bovair, S. (1984) The role of a mental model in learning to operate a device. *Cognitive Science* 8(3): 255–273. doi:10.1016/S0364-0213(84)80003-8.
Loborik, J., Corry, N., Rayner, J., Darling, A., Dougherty, K., John, D., and Beecroft, S. (2015) *Doctor Who: The Visual Dictionary.* London: DK Publishing.
Lofficier, J.M. (1992) *The Universal Databank.* London: W.H. Allen & Co.
Lofficier, J.M. (n.d.) Foreword. *Doctor Who: The Universal Databank.* www.lofficier.com/dwfrwrd.html (accessed 29 January 2020).
Mann, G., and Richards, J. (2016) *Doctor Who: The Whoniverse.* New York: Harper Design.
Mann, G., Richards, J., and Scott, C. (2017) *Doctor Who: Dalek.* London: BBC Books.
McGonigal, J. (2003) "This is not a game": Immersive aesthetics and collective play. In: Miles, A. (ed.) *Melbourne DAC 2003 Streamingworlds Conference Proceedings.* janemcgonigal.files.wordpress.com/2010/12/mcgonigal-jane-this-is-not-a-game.pdf.
Mittell, J. (2007) Lost in a great story. *Just TV.* justtv.wordpress.com/2007/10/23/lost-in-a-great-story (accessed November 10, 2019).
Nation, T. (1965) *The Dalek Pocketbook and Space-Travellers Guide.* London: Panther Books.

Nation, T. (1978) *Terry Nation's Dalek Annual 1978*. Manchester: World Distributors.
Nation, T. (1979) *Terry Nation's Dalek Annual 1979*. Manchester: World Distributors.
Nation, T., and Ashton, B. (1966) *The Dalek Outer Space Book*. London: Souvenir Press.
No Author (1962) *Dan Dare's Space Annual 1963*. London: Longacre Press Limited.
No Author (1985) *The Dalek Problem: A Symposium*. Chicago: FASA.
No Author (1992) *Doctor Who Yearbook 1993*. London: Marvel.
No Author (2002) *The Dalek Survival Guide*. London: BBC Books.
No Author (2016) *Doctor Who: A History of Humankind: The Doctor's Official Guide*. London: BBC Children's Books.
O'Day, A. (2017) A rude awakening: Metafiction in Eric Pringle's "The Awakening." In: Fleiner, C., and October, D. (eds.) *Doctor Who and History: Critical Essays on Imagining the Past*. Jefferson: McFarland, pp: 92–101.
Perryman, N. (2008) Doctor Who and the convergence of media: A case study in "transmedia storytelling." *Convergence: The International Journal of Research into New Media Technologies* 14(1): 21–39. doi:10.1177/1354856507084417.
Rehak, B. (2016) Transmedia space battles: Reference materials and miniatures wargames in 1970s Star Trek fandom. *Science Fiction Film & Television 9(3)*: 325–345. doi:10.3828/sfftv.2016.9.9.
Rehak, B. (2018) *More Than Meets the Eye: Special Effects and the Fantastic Transmedia Franchise*. New York: New York University Press.
Richards, R. (2017) *The Dangerous Book of Monsters*. London Penguin Books.
Ryan, M.L. (2017) The aesthetics of proliferation. In: Boni, M. (ed.) *World Building: Transmedia, Fans, Industries*. Amsterdam: Amsterdam University Press, pp. 31–46.
Saler, M. (2012) *As If: Modern Enchantment and the Literary PreHistory of Virtual Reality*. New York: Oxford University Press.
Sillars, S. (1995) *Visualisation in Popular Fiction 1860–1960: Graphic Narratives, Fictional Images*. New York: Routledge.
Standing, L. (1985) *The Doctor Who Illustrated A-Z*. London: W.H. Allen & Co.
Suvin, D. (1979) *Metamorphoses of Science Fiction: On the Poetics and History of a Literary Genre*. New Haven: Yale University Press.
Tavares, S. (2015) Paratextual prometheus. Digital paratexts on YouTube, Vimeo and Prometheus transmedia campaign. *International Journal of Transmedia Literacy* 1(1): 175–195. doi:10.7358/ijtl-2015-001-tava.
Turner, A.W. (2011) *The Man Who Invented the Daleks: The Strange Worlds of Terry Nation*. London: Aurum Press.
Whitaker, D., and Nation, T. (1964) *The Dalek Book*. London: Panther Books.
Whitaker, D., and Nation, T. (1965) *The Dalek World*. London: Souvenir Press.
Wolf, M.J.P. (2012) *Building Imaginary Worlds: The Theory and History of Subcreation*. New York: Routledge.

The Science of *Doctor Who*

MARK ERICKSON

Introduction

Theoretical physicist Lawrence Krauss published *The Physics of Star Trek* in 1995, the first in a long series of *The Science of [Insert SciFi Title Here]* texts, itself a subset of a very prominent and persistent genre: popular science (O'Keeffe, 2017). Krauss' book was followed in fairly rapid succession by, among others, *The Science of the X-Files* (White 1998), *The Science of Jurassic Park* (De Salle and Lindley 1997), *The Science of the X-Files* (Cavelos 1998b), *The Science of Star Wars* (Cavelos 1998a), *The Science of Discworld* (Pratchett et al. 1999), *The Real Science Behind the X-Files* (Simon 1999), *The Science of Superheroes* (Gresh and Weinberg 2002), *The Science of Middle Earth* (Gee 2004), *The Science of Harry Potter* (Highfield 2003).... Late arrivals to this list of books were two *Doctor Who*-themed "science of" texts: Michael White's *A Teaspoon and an Open Mind; The Science of Doctor Who* (2005) and Paul Parsons' *The Science of Doctor Who* (2006), both arriving on the back of the dramatic resurgence of popularity of the *Doctor Who* TV series following its relaunch on BBC Television in 2005. Parsons' book was also offered as a free gift for new subscribers to *BBC Focus* magazine in 2006, and issue #162 ran an eight-page feature on the book (Parsons was editor of *BBC Focus* in 2006). The "Science of..." trope clearly retained its popularity, at least in the eyes of the BBC, who celebrated the 50th anniversary of *Doctor Who* with an hour-long lecture titled *The Science of Doctor Who* delivered by Professor Brian Cox and broadcast on BBC2 (Thursday 14 November 2013).

The paratext—a text that runs parallel or at a tangent to an original text (Gray 2010; Hills 2015)—serves to complement the original text in a number of ways. For Matt Hills, the *Doctor Who* paratexts that emerged around the 50th anniversary served to not only deepen the experience of fans with respect to a beloved TV series, they also served to provide legitimacy for the BBC and shaped the way that viewers understood the TV program (Hills 2015). While this is certainly the case for many of the *Doctor Who* paratexts that emerged at the time of the 50th anniversary—TV specials, trailers, docudramas about the creation of the series—the *Science of Doctor Who* paratexts do something quite different. Rather than deepening the fan experience of the TV series through extension and fleshing out of plot lines, revealing how the series is made, or presenting the TV series in different media, these paratexts take the reader to a completely different place; away from the fictional universe and into, literally, the universe itself as understood by formal science.[1] They do this by invoking and articulating a strongly normative and didactic account of formal science and, in doing so, serve to accentuate our sense that formal science is superior to other forms of knowledge and other practices of inquiry. In this essay, I will look at how these *Science of Doctor Who* paratexts, while clearly having some connection to other *Doctor Who* paratexts, are better understood as a subset of the much larger genre of popular science texts that remain very popular in contemporary culture. We can see these strong resemblances through a close reading of *The Science of Doctor Who* texts.

Popular Science

Popular science is formal science that has, through a process of translation, become popularized and presented, through various media, to a wide and amateur audience. Where some commentators (e.g., Sismondo, 2009) see popular science as largely homogenous and an enterprise carried out predominantly by journalists, I think we can discern quite considerable variation in this genre in terms of media, authors and content. We can also note some quite specific functions that popular science serves. Finally, across media and authors, we can find some core shared themes that are across these texts. Before looking in detail at *The Science of Doctor Who* paratexts some examination of the genre of popular science is necessary.

Exoteric and Esoteric Popular Science

Given that popular science is such a large genre (it often has its own section in a bookshop, and prominent new titles emerge very frequently) it is unsurprising that we find a wide range of texts and authors. One of the first things that we can notice about popular science texts is that they are grouped into two large camps with some overlapping aspects: texts that want to popularize formal science, and texts that want to "save" the lay public from the dangers of pseudoscience, quackery and false scientific analysis (Erickson 2015: 166). In the latter camp there are some venerable authors and texts: Carl Sagan's *The Demon-Haunted World: Science as a Candle in the Dark* (1996) is, perhaps, the best-known example, but this sub-genre began in the 1950s with Martin Gardner's *Fads and Fallacies in the Name of Science* (1957) and continues well into this century with, for example, the work of Massimo Pigliucci (2010) and Michael Shermer (2016).

In the other large grouping, where texts are presenting formal scientific knowledge in popular formats, we can note significant differences in terms of content between texts aimed at very general audiences (e.g., Bill Bryson's *The Body: A Guide for Occupants*, 2019) and those aimed at more specialist audiences (e.g., texts by Stephen Hawking [1988]; Hawking and Mlodinow 2011). Here we find texts spread on a continuum, one that largely parallels the institutional location of their authors. Sociologist of science Ludwik Fleck introduced the concept of "thought communities" in his ground-breaking book *Genesis and Development of a Scientific Fact* (1935) to describe how knowledge emerges in different forms according to shared social activity. While we are all members of a number of exoteric thought communities, where we share broadly similar styles of thought (neighborhoods, political parties, faith groups), formal scientists are members of, usually, a single esoteric thought community where the style of thought is focused on shared knowledge of a very specific set of practices and objects of inquiry (e.g., a molecular biology laboratory in a university with connections to other molecular biology laboratories at other academic and research institutions). Popular science texts that are produced by authors located within esoteric thought communities have, in general, a very different feel and style to them, and are directed at different audiences, than those that emerge from exoteric locations (such as those produced by science journalists or celebrities). We can see popular science texts arrayed along a continuum between exoteric and esoteric depending on content and audience. A couple of examples will serve to illustrate this.

Janna Levin's *Black Hole Blues and Other Songs from Space* (2016) is an account of the origins and development of the LIGO project to detect gravitational waves. Based on first-hand accounts with key participants the book offers a lively narrative of the actions and interactions of the main protagonists in the discovery of gravitational waves. Although written largely in the first person, and with extensive quotations from interviews and interactions, the book presents an esoteric account where considerable knowledge of, and interest in, theoretical physics and cosmology is something of a prerequisite. This is an esoteric popular science book aimed at an audience with a good degree of familiarity with the topic by a member of a shared esoteric thought community (as Levin notes herself on first meeting Rainer Weiss, the inventor of the "ifo," the interferometer that is the key component of the LIGO experiment, see LIGO n.d.):

> We skip conventional social openers and speak with familiarity, although this is our first meeting, as though we've known each other for as long as imaginable, the shared experience of our scientific community outweighing a shared hometown or even generation. We lean back in mismatched chairs, our feet propped up on a single stool [Levin 2016: 7].

Towards the other end of the continuum we find texts written by, often, science journalists, with a didactic feel and simplified, and chattily explained, contents. Science popularizer, actor and comedian (and physics Ph.D.) Ben Miller's recent book *The Aliens Are Coming: The Extraordinary Science Behind Our Search for Life in the Universe* (2017) is a witty and light-hearted overview of recent attempts to conceptualize, find and contact alien life outside our solar system. Miller's book is, typically of this style of text, full of praise for formal science (it is, for example, "extraordinary") and posits science against the "other" of foolishness, ignorance, credulity. Noting that our attempts to find alien life have as yet yielded nil results, Miller talks directly to the reader:

> But before you throw this book down in a pique of anti-science disgust and head for the Mind, Body and Spirit section, stop. Because as is so often the case, the real science is so much more interesting than the non-scientific stuff. While alien autopsies grab the headlines, thousands of scientists—real, hardworking, peer-reviewed, genuinely qualified scientists—are slowly inching closer to the real thing [Miller 2017: 6].

Miller's placing of "real" science in direct opposition to the foolishness of "Mind, Body and Spirit," i.e., belief, is a very standard move throughout the popular science genre. But Miller is also placing "real" science in opposition to science fiction, to speculation and is making a further claim: that science fiction cannot be scientific. This trope—isolating "real" science from the rest of society and setting it up as superior precisely due to this separation from society, is almost *de rigueur* in popular science texts.

The Role of Popular Science in Contemporary Culture

Popular science texts are, as the name implies, popular and they inform, educate and entertain a great many people. But, regardless of their position on the exoteric-esoteric continuum, as they do this they invoke an "other": foolishness and ignorance. We can see this being implied quite clearly in some of these texts' subtitles: Sagan's "candle in the dark," Shermer's "where sense meets nonsense," Park's "the road from foolishness to fraud" (2000) and "belief in the age of science" (2010). Throughout the genre science is routinely portrayed as the rational opposite of organized foolishness and credulity, the popular science definition of religion and/or faith. In contrast to foolishness, popular science texts present a very strong, and very widely accepted, narrative of scientific progress that leads to a triumphal outcome—complete knowledge of the natural world. *En route*, popular science presents science as being a unitary phenomenon, with all formal science practices, scientists and institutions subsumed into one shared collective enterprise and institution: the scientific community which is united in its shared purpose and method. On closer examination we can see that this collective is actually arranged into a hierarchy with physics at the apex of both knowledge and practice, a view that has held sway throughout the twentieth and twenty-first centuries and is exemplified in the attitude and the project of the Vienna Circle, neatly summed up by philosopher of science Ian Hacking:

> They disagreed about much, but only because they agreed on the basics. They thought that the natural sciences are terrific, and that physics is the best. It exemplifies human rationality. It would be nice to have a criterion to distinguish such good science from bad nonsense or ill-formed speculation [Hacking 1983: 3].

This description could equally be applied to a great many popular science authors.

There are some additional roles that popular science performs. As David Bell notes, the emergence and expansion of popular science in its various forms served to harden the distinction between amateur and professional (Bell 2006) with professional scientists engaging with quite different texts from amateurs and the lay public. This, we can see, also underpins the construction and articulation of the other of science: it is only through the adoption of a scientific worldview that individuals can be competent, legitimate commentators on matters technical or environmental.

Tropes and Themes in Popular Science

Across the continuum of popular science texts we can find a core set of shared themes and tropes that characterize this genre and four, in particular, stand out: history, triumphalism/superiority, collectivization and gender.

Popular science texts, almost universally, state or imply a continuous and triumphal history of science, a history of significant discoveries and breakthroughs made across the centuries. This is the foundation upon which our contemporary scientific worldview is constructed: science was great in the past and is getting better all the time in a progressive and stepwise manner. Across large areas of history of science (e.g., Kuhn

1962), philosophy of science (e.g., Feyerabend 1978) and science and technology studies (STS) (e.g., Latour and Woolgar 1979) this account has been thoroughly called into question, yet our dominant story persists and is sustained by these popular science accounts.

The successful history of science underpins the second theme, where science is seen as a form of knowledge that is superior to other forms of knowledge. We saw above how popular science constructs the "other" of science as foolishness, ignorance, stupidity and credulity (Erickson 2015: 166). This dialectical choice—science or ignorance—is a persistent trope in all forms of popular science texts and presents scientific knowledge, the scientific worldview and scientism as the mode of understanding that is simply better than all others (Bensaude-Vincent 2009). Stephen Hawking expressed this clearly as he dispatched competing cosmological perspectives in his final popular science text *The Grand Design*:

> How can we understand the world in which we find ourselves? What is the nature of reality? Where did all this come from? ... Traditionally these are questions for philosophy, but philosophy is dead. Philosophy has not kept up with modern developments in science, particularly physics. Scientists have become the bearers of the torch of discovery in our quest for knowledge [Hawking and Mlodinow 2010: 13].

Of course scientism is not limited to non-fiction, and some eras of *Doctor Who*, in particular the Pertwee and Tom Baker eras of the 1970s, have promoted similarly scientistic worldviews (Orthia 2010, 2011; see also Orthia and de Kauwe this volume). As Lindy Orthia (2011) details, the "gothic horror" era during the first three years of Tom Baker's tenure as the Doctor was replete with storylines in which a scientific explanation for phenomena triumphed over supernatural or religious beliefs. Viewed from this perspective, *Doctor Who* itself could be considered a popular science text when it has included snippets of scientific "fact" and promoted scientistic ideologies.

This takes us to the third shared theme across popular science, and further: the collectivization of all formal scientists into a single community with a single shared method. Again, against large swathes of STS, history and philosophy of science research and analysis to the contrary, the collectivization of the very diverse disciplines and participants into a single grouping—scientists or "the scientific community"—is the standard trope for almost all popular science texts, replicating the discourse of formal scientists themselves (Erickson 2015). Again this conceptualization of scientists appears frequently in *Doctor Who*, for example in "Planet of Evil" (1975) when the Doctor tells Professor Sorenson, who creates chaos and death trying to harness the power of antimatter: "You and I are scientists, Professor. We buy our privilege to experiment at the cost of total responsibility."

The fourth theme is that of gender and the gendered discourse of formal science and scientific institutions. Looking again at our first theme—the triumphal history of science—we can see that this trope itself is gendered. The discoveries that pile up stepwise to make our triumphal history were made by men, and where women were involved in these, they took only supporting or marginal roles. Examples are legion; from John Gribbin's *Science: A History* (2003) which lists the great men of science in chronological order to show "the development of science in essentially incremental, step-by-step terms" (2003: 614) to Tim Radford's 2018 *The Consolations of Physics: Why the Wonders of the Universe Can Make You Happy* which includes only a single source written by a woman (Jacquetta Hawkes' *A Land* written in 1951). The message being conveyed is that science is an enterprise by and for men. Once again, this is in very stark contrast to the research carried out by historians of science and STS researchers. Patricia Fara's *Science: A Four Thousand Year History* (2009) is the best and most comprehensive of many historical accounts uncovering and highlighting the role of women in the history of science, and showing the systematic exclusion and discrimination that women in science have faced in the past, and still face today. Donna Haraway (1997, 2016), Sandra Harding (1986, 1991) and Hilary Rose (1994), foremost amongst many others, have presented comprehensive arguments against the androcentric epistemological assumptions of formal science, and shown that such starting points lead to poorer scientific analyses.

The following sections look at how our *The Science of Doctor Who* paratexts deploy these themes, and then go on to consider some implications for this. But first, a quick summary and characterization of these popular science paratexts is in order.

The Science of Doctor Who Texts as Popular Science

As noted above we can see popular science texts as falling into one or other of two large camps. The three paratexts considered here—Brian Cox's BBC Royal Institution lecture titled "The Science of *Doctor Who*," Michael White's 2005 book *A Teaspoon and an Open Mind: the Science of Doctor Who*, and Paul Parsons' 2006 book *The Science of Doctor Who*—are all in the "popularization" of formal science group: their aim is to present complex scientific theories and ideas to a "lay" audience, and they use the vehicle of *Doctor Who* as a means of drawing in an audience and structuring their text. This is done with varying degrees of engagement with *Doctor Who*; as Hills notes, the BBC2 Brian Cox program was:

an old-fashioned lecture combined with drama inserts filmed on the TARDIS set—in these, Cox interacted with the eleventh Doctor, Matt Smith. Connections made to *Doctor Who* in the lecture material were somewhat underdeveloped, and the programme awkwardly layered popular science on to its *Doctor Who* "hook" [Hills 2015: 41].

In contrast, Parsons' book has a very active engagement with the plots, technologies and characters of the *Doctor Who* "mythos" (Parsons 2006: xv); Moira O'Keeffe (2017: 34) describes the book as "respectful of both the show and the science." There are 34 short chapters in the book, each prefaced by an epigraph from a *Doctor Who* character, all of which use some element of *Doctor Who* as a springboard into formal science. The book also includes a list of episodes and a list of the (at that point) ten actors who had played the Doctor.

However, White's book is considerably different; while using some elements of *Doctor Who* as starting points into the "real" science, for example noting that the Doctor is a Time Lord then going on to consider the science of time travel, this paratext has remarkably little engagement with the text it is supposedly about. This is particularly clear in Chapter 3: "Bring on the Daleks" which makes no mention whatsoever of Daleks, the Doctor, Time Lords or anything else that a *Doctor Who* audience might expect (the chapter is 20 pages long). This contradicts one of the key criteria O'Keeffe (2017) identifies as essential for a paratext's author to create an authentic connection with fan-readers: demonstrated knowledge of the original text. White, like Parsons, precedes each chapter with an epigraph but these are all taken from literary history (Marcus Aurelius, Sherlock Holmes, Isaac Asimov, Mark Twain, and so on). In the book White offers speculation, coupled to contemporary scientific theories, regarding the development of artificial intelligence, the existence of life on other planets, human consciousness, extending human lifespans (amongst other things).

Considering these on our spectrum of popular science texts, the BBC2 program is much closer to the esoteric end, with its relentless focus on formal science, evidenced by Cox's delivery of hard science content, including drawing space time diagrams on a chalkboard and reproducing science experiments live on stage. Slightly less esoteric is White's text, which presents "hard" science detail in the form of descriptions of current formal science theories, such as an extended description of Heisenberg's Uncertainty Principle (pages 125–131). Much more exoteric is Parsons' text which presents formal science in very truncated, "bite sized" pieces (e.g., explaining genes and DNA in two sentences on page 7) and is, at times, slightly irreverent. We will now look at the four core tropes of popular science as they appear in these paratexts.

History in "The Science of Doctor Who" Paratexts

Brian Cox opens his lecture with the following words:

> If I could borrow the TARDIS just for one day, of all the places I would travel through space and time I'd choose here on the 27th of December 1860. On that day Michael Faraday stood on this stage and delivered his Royal Institution Christmas Lecture on the chemical history of the candle. [...] Faraday was one of the greatest scientists in the world. He laid the foundations of our modern understanding of electricity and magnetism [4'12"].

As noted earlier, popular science invokes a strong, progressive and triumphal history of science, characterized by (mostly) men of genius making great discoveries, in this case Michael Faraday. Cox draws a direct connection between Faraday's work and our contemporary techno-scientific world, implying that our world would be considerably different had Faraday not existed. Faraday's work by Cox's account is foundational, drawing our attention to the building block model of the history of science where men like Newton and Faraday stand on the shoulders of giants and deliver their singular and awesome vision. Perhaps surprisingly, this chronological and stepwise approach to history is remarkably similar to that presented in academic histories of science, albeit including much greater detail and depth. Academic histories of science almost invariably present an account of science that is entirely separate from the rest of society, with only mentions of non-science aspects of history being made when absolutely necessary (Erickson 2010).

These histories of science convey not just trajectory (ever upwards) but also greatness, genius and the superior vision of the project of science. White's book provides a good example of stepwise, genius-driven progression (170): "Imagination has served us well so far, for without it, Einstein, Newton, Darwin and all the other great pioneers of science would have failed to tease their own tiny bit of magic from the jealous grasp of Nature." Orthia (2010: 60–62) notes that White's book also endorses a stagist view of human history in which western science and technology are used as the benchmark of a society's level of evolution. This further reifies the notion of the triumphal march of science (and implicitly western culture) towards progress and greatness.

In contrast, Parsons' book is much more light-hearted, as he notes in the Epilogue "If I did manage to educate anyone along the way, then I sincerely apologise" (317), but still makes reference to some of the standard tropes of the history of science (e.g., Einstein working in a Swiss patent office [32], Newton's 1687 theory of gravity [34]). Here we see the necessity of including signifiers of the greatness of the history of science and its continuing triumph, in this case in the form of two of its most renowned practitioners, Einstein and Newton. The triumph of the history of science becomes, in popular science, a theme on its own and our *Science of Doctor Who* paratexts are no exception to this.

Triumphalism/Superiority in "The Science of Doctor Who" Paratexts

Triumphalism and the superiority of the scientific method and worldview are visible in almost all popular science texts, and also in these paratexts. For example, Brian Cox, no stranger to the use of superlatives, presents a conflation of the progressive history of science and the triumph of the scientific vision, in his description of the consequences of Faraday's work: "The elegant answer was provided by the great Scottish physicist James Clerk Maxwell" (14'15"). Maxwell is introduced as Cox explains that the speed of light does not change and this realization is part of "Maxwell's genius" (14'31"). Cox then goes on to tell us again that the speed of light is never changing, being always the same regardless of where it is measured and by whom: "It is always, in modern units, precisely 299,792,458 meters per second" (16'15")—and Cox repeats this four times, for slightly comic effect, between 15'03" and 17'43". It is, unfortunately, not a true statement but, surprisingly for a professor of physics, Cox doesn't point out that the speed of light varies according to the medium it is being propagated in. The constant "c"—299,792,458 meters per second—is the speed of light *in a vacuum*, but light travels slower in other media and if you measure the speed of light in air, water and vacuum each such measurement will yield a different result. It is hard to imagine that Cox would allow his undergraduate physics students to get away with such poor definitional work, but exaggeration for effect (for that is what this is) is quite acceptable for a "lay" audience: the speed of light never changes—isn't that amazing!

Science has to be described in very positive terms in popular science texts. Here is Michael White:

> The great thing about science is that it is always changing and growing. Along with the fact that science is based on experiment and analysis, it is this ability to develop that really separates it from *mere* belief systems. But, even though science evolves, every good scientist goes with the flow [x—my emphasis].

Here, White is not only pointing to how good science is, but is also deploying that very common popular science trope of positing science against an "other" that is inferior, in this case "belief." White, throughout the book, makes the case for the standard, normative account of science, one that is at odds with even mainstream and thoroughly pro-science philosophers of science. For example, in describing Einstein's theory of relativity and the impossibility of time travel, White says:

> [T]his is not a concocted story designed merely to spoil our fun. These are universal, irrefutable facts, and the theory of relativity has been proved correct by a century of experiments. Indeed, it is rather ridiculous to still call the theory of relativity a theory at all [64].

This is quite at odds with, for example, the philosophy of Karl Popper, which notes that no scientific theory can ever be proved outright and that, indeed, is what gives scientific knowledge its strength (Popper 2002).

Even Parsons' light-hearted and at times irreverent book feels the need to point to the wonderfulness of science: "[T]his is not a book of anorak-clad, nasal-toned pedantry. Rather, it's a gathering of amazing possibilities" (xv). Parsons' use of biological determinism to explain social phenomena such as altruism (9–10) further contributes to an underlying flavor of scientism in this book (Orthia 2010).

Collectivization in "The Science of Doctor Who" Paratexts

An important aspect of the standard and normative account of science is to portray "science" as a unified project: unified by method, goals and ethos. While this unification is central to the standard account (Merton 1957; Hagstrom 1965; Kuhn 1962), it is at odds with many STS and other social scientific and philosophy accounts of how formal science is done and understood by its practitioners. Lumping together *all* practitioners of formal science—from paleontologists to physicists—elides very significant differences between disciplines and, according to Steve Fuller, is really an artificial construct that emerged in the 1950s to serve a political purpose, to escape calls for public accountability (Fuller 2000: 209). Popular science texts persist in this "collectivization" of all scientists and disciplines, and this also serves to set science as an activity and a form of knowledge as somehow apart from the rest of society. Cox, in his lecture, does this through identifying the unified project of science in an historical context (5'22"): "Another of the great scientific ghosts that haunt this place on Albemarle Street, Humphry Davy, the charismatic professor of chemistry at the Royal Institution, and a passionate believer in the power and possibility of science." He goes on to note that Davy's and others' work has led to "[t]he great discoveries that have shaped our scientific civilization" (5'55").

Michael White's text deploys frequent collectivizations, bringing different disciplines under a single title. For example, in discussing human genetics (xi): "Until very recently it was believed that the only useful parts of the genome were the genes themselves. But scientists have now found that what was originally called 'junk' DNA [...] may be vitally important."

And in similar style, Parsons also collectivizes frequently. For example (128): "According to scientists, breathing life into plastic isn't all that difficult. Researchers at the University of Sheffield are studying plastics that can change shape in a precisely controlled way."

The question of "which scientists" is never asked, nor even hinted at. We "know," from the strong and dominant narrative of science in our society, that scientists are all part of one large body, and that this is separate from the non-scientists/non-scientific world (Erickson 2015).

Gender in "The Science of Doctor Who" Paratexts

Cox asks for volunteers from the audience, a mix of celebrities and the general public, to help with the experiments he has chosen to illustrate his lecture. There looks to be a pretty even gender, and age, distribution across this audience, but Cox's "assistants"—science publicizer Dallas Campbell, physicist Jim Al-Khalili, actor Charles Dance and comedian Rufus Hound—are all men. As we proceed through the lecture, Cox introduces a number of important scientists, but, again, these are all men (Faraday, Einstein, Fermi, Davy, Maxwell, Kepler). And our two stars—Brian Cox himself and Matt Smith as the Doctor—are both men. Through the entire lecture, the only female faces we see are sitting in the audience, and there are no female voices heard. And without any sense of irony the program finishes back on the TARDIS set with Matt Smith (the Eleventh Doctor) telling Brian Cox that there is an "ordinary kid" in his audience, but *she* will, after being inspired by Cox's lecture, stop being ordinary as "she grows up to be extraordinary, a woman who changes the world. And all she needed was a nudge from you, eh? Today, right now. No pressure" (58'56").

There are two women mentioned in White's book. The first is renowned astronomer Dame Jocelyn Bell Burnell, who is name-checked in one sentence: "In 1967, the British astronomer Jocelyn Bell discovered the first pulsar" (15). The second appears in discussing the formation of Team Encounter, an eclectic group of people interested in making contact with alien life: "This group of enthusiasts, which includes exobiologists, the first female astronaut, Sally Ride,[2] and the rock musician Greg Lake..." (39). Ride's achievements and distinguished post–NASA career (she was, amongst other things, professor of physics at UC San Diego) are not mentioned. One other woman, Madame Helena Blavatsky, the founder of the Theosophical movement and attractor of many critics who saw her as a fraudulent charlatan, is included in the book; in contrast to Bell Burnell and Ride, Blavatsky is discussed at some length across six paragraphs (111–12).

Unlike either Cox' or White's texts, Parsons' book does address issues of gender and sex, and does this with direct reference to *Doctor Who*. In the section of chapter 6 titled "Earth girls are easy" Parsons notes "Rose Tyler fleeing the Autons probably isn't what fans are referring to when they talk about the show's bouncing wit" (64). This kind of casual sexism is endemic in UK society, of course, but in this context feels particularly inappropriate. Making STEM subjects attractive to women and girls is an imperative that is noted by almost all formal scientific institutions (see for example CASE, 2014), almost all of which recognize that sexism and sexist attitudes inside scientific communities and the gendered ways that science is represented in public is a major problem. Here, for example, is Paul Hardaker, CEO of the Institute of Physics:

> In fact, generations of innovative, talented and brilliant girls are being led to believe they can't be engineers, scientists, programmers or technicians. Sometimes it is idle comments that have the deepest effect in discouraging girls from taking physics to a higher level [Hardaker, 2018].

Conclusion

We have looked at three quite different *Doctor Who* paratexts. Cox's lecture teaches us about physics, but avoids much contact with *Doctor Who*. White's book provides a space for the author's speculations about aspects of formal science they are interested in and, although it avoids much contact with *Doctor Who*, does make some passing reference to the Doctors, their companions and adversaries. Parsons' book, the most exoteric of the three, actively engages with the *Doctor Who* mythos in a lively way, but also provides considerable information about contemporary science theories and areas of interest. Of the three, I think it is only Parsons who is really producing a paratext—a text that extends our engagement with the original text—and then only as a starting point to take us away from science fiction and into the world of "real" science. So while our three texts look like paratexts, adopting the branding and some content of *Doctor Who*, they are much better understood as representatives of a different genre—popular science—and could perhaps be better described as "parasitical texts."

While relatively critical of *Doctor Who*'s representations of science, Orthia (2010) notes that fiction can sometimes encourage equality in the science workplace and contribute to the democratization of science, and we can think of *Doctor Who* as an opportunity, a tool, to open the way for students and indeed teachers to articulate our thoughts and feelings about what science is and the role we would like to see it play in our societies, cultures and individual lives [294].

Nonetheless, the irony is that popular science, as this analysis of these *Doctor Who* paratexts shows, generally works in the opposite direction, reinforcing gender stereotypes about science and hardening the barrier between formal science and the wider public.

Popular science educates, entertains and informs a great many people: it is, after all, "popular." However, it also serves as a vehicle that articulates and consolidates the standard account of science and a scientistic worldview. Philosopher Mary Midgley provides a stark definition of scientism: "[T]he ambition to take over the whole of human knowledge for physics and chemistry" (Midgley 2010: 92). Popular science serves to reinforce scientism, and scientism provides the metanarrative to popular science (Erickson 2015). In its most simple form, scientism suggests that explanations for the world around us (be that the "natural" world or the social world or the personal world of the mind) should be based on scientific principles rather than religion, superstition, guesswork or, significantly, "non-scientific" social science or philosophy. It articulates a coherent, linear and progressive history for science, one that implies a completion of the project by achieving the goal of total knowledge. It represents scientific method and scientific knowledge as unified and as inherently superior to other forms of knowledge and personal experience. It presents science as profoundly androcentric and articulates this in sexist and gendered ways. It is this final point that is, perhaps, the most ironic. When even the Doctor can change gender (see Stack this volume), why is it that formal science and its cheerleader popular science cannot address its fundamentally gendered nature?

Notes

1. I deliberately use the term "formal science" rather than "natural science," where formal science is the practices and knowledges that are carried out and articulated in accredited scientific institutions (universities, research laboratories, etc.) and become formalized through institutional acceptance. This is often labelled "natural science" but this latter term implies a privileged and direct connection to "nature" that is somehow denied to the non-scientist. See Erickson 2015.

2. I can find no reference to Sally Ride's participation in this project in any of her many biographies and obituaries.

References

Bell, D. (2006) *Science, Technology and Culture*. Maidenhead: Open University Press.
Bensaude-Vincent, B. (2009) A historical perspective on science and its "others." *Isis* 100(2): 359–386. doi:10.1086/599547.

Bryson, B. (2019) *The Body: A Guide for Occupants*. London: Doubleday.
CASE (Campaign for Science and Engineering) (2014) *Improving Diversity in STEM*. London: CASE.
Cavelos, J. (1998a) *The Science of Star Wars: An Astrophysicist's Independent Examination of Space Travel, Aliens, Planets, and Robots as Portrayed in the Star Wars Films and Books*. New York: Saint Martin's Press.
Cavelos, J. (1998b) *The Science of The X-files*. New York: Berkley Boulevard Books.
DeSalle, R., and Lindley, D. (1997) *The Science of Jurassic Park and The Lost World or, How to Build a Dinosaur*. London: HarperCollins.
Erickson, M. (2010) Why should, I read histories of science? *History of the Human Sciences* 23(2): 68–91. doi:10.1177/0952695110372022.
Erickson, M. (2015) *Science, Culture and Society: Understanding science in the twenty-first century*, 2nd edition. Cambridge: Polity.
Fara, P. (2009) *Science: A Four Thousand Year History*. Oxford: Oxford University Press.
Feyerabend, P. (1978) *Against Method*. London: Verso.
Fleck, L. ([1935] 1979) *Genesis and Development of a Scientific Fact*. Chicago: University of Chicago Press.
Fuller, S. (2000) *The Governance of Science: Ideology and the Future of the Open Society*. Buckingham: Open University Press.
Gardner, M. (1957) *Fads and Fallacies in the Name of Science*. New York: Dover Publications Inc.
Gee, H. (2004) *The Science of Middle-earth*. Cold Spring Harbor: Cold Spring Press.
Gray, J. (2010) *Show Sold Separately: Promos, Spoilers, and Other Media Paratexts*. New York: New York University Press.
Gresh, L.H., and Weinberg, R.E. (2002) *The Science of Superheroes*. Hoboken: J. Wiley.
Gribbin, J. (2003) *Science: A history 1543–2001*. London: Penguin.
Hacking, I. (1983) *Representing and Intervening: Introductory Topics in the Philosophy of Natural Science*. Cambridge: Cambridge University Press.
Hagstrom, W.O. (1965) *The Scientific Community*. New York: Basic Books.
Haraway, D. (1997) *Modest-Witness@Second-Millennium.FemaleMan-Meets- OncoMouse: Feminism and Technoscience*. London: Routledge.
Haraway, D.J. (2016) *Staying with the Trouble: Making Kin in the Chthulucene*. Durham: Duke University Press.
Hardaker, P. (2018, October 9) IOP CEO Paul Hardaker speaks out on gender inequality in science, *Institute of Physics*. www.iop.org/news/18/october/page_72112.html (accessed 29 January 2020).
Harding, S. (1986) *The Science Question in Feminism*. Ithaca: Cornell University Press.
Harding, S. (1991) *Whose Science? Whose Knowledge? Thinking from Women's Lives*. Milton Keynes: Open University Press.
Hawking, S.W. (1988) *A Brief History of Time: From the Big Bang to Black Holes*. New York: Bantam Books.
Hawking, S.W., and Mlodinow, L. (2011) *The Grand Design*. London: Bantam Press.
Highfield, J.R.L. (2003) *The Science of Harry Potter: How Magic Really Works*. London: Headline.
Hills, M. (2015) *Doctor Who: The Unfolding Event—Marketing, Merchandising and Mediatizing a Brand Anniversary*. London: Palgrave Macmillan.
Krauss, L.M. (1995) *The Physics of Star Trek*. London: HarperCollins
Kuhn, T.S. (1962) *The Structure of Scientific Revolutions*. Chicago: University of Chicago Press.
Latour, B., and Woolgar, S. (1979) *Laboratory Life: The Social Construction of Scientific Facts*. London: Sage.
Levin, J. (2016) *Black Hole Blues and Other Songs from Outer Space*. London: The Bodley Head.
LIGO (Laser Interferometer Gravitational-Wave Observatory) (n.d.) LIGO's Interferometer. *LIGO Caltech*. https://www.ligo.caltech.edu/page/ligos-ifo (accessed 17 October 2019).
Merton, R.K. (1957) *Social Theory and Social Structure*. New York: Free Press.
Midgley, M. (2010) *The Solitary Self: Darwin and the Selfish Gene*. Durham: Acumen.

Miller, B. (2017) *The Aliens Are Coming: The Extraordinary Science Behind Our Search for Life in the Universe*. London: Sphere.

O'Keeffe, M. (2017) Riding the wave: Science fiction media fandom and informal science education. *Journal of Science Fiction* 1(3): 24–39.

Orthia, L.A. (2010) *Enlightenment was the Choice: Doctor Who and the Democratisation of Science*. Ph.D. Thesis, The Australian National University.

Orthia, L.A. (2011) Antirationalist critique or fifth column of scientism? Challenges from Doctor Who to the mad scientist trope. *Public Understanding of Science* 20(4): 525–542. doi:10.1177/0963662509355899

Park, R. (2000) *Voodoo Science: The Road from Foolishness to Fraud*. Oxford: Oxford University Press.

Park, R.L. (2010) *Superstition: Belief in the Age of Science*. Princeton: Princeton University Press.

Parsons, P. (2006) *The Science of Doctor Who*. Thriplow: Icon Books.

Pigliucci, M. (2010) *Nonsense on Stilts: How to Tell Science from Bunk*. Chicago: University of Chicago Press.

Popper, K. (2002) *The Logic of Scientific Discovery*. London: Routledge Classics.

Pratchett, T., Stewart, I., and Cohen, J. (1999) *The Science of Discworld*. London: Ebury.

Radford, T. (2018) *The Consolations of Physics: Why the Wonders of the Universe Can Make You Happy*. London: Sceptre.

Rose, H. (1994) *Love, Power and Knowledge: Towards a Feminist Transformation of the Sciences*. Cambridge: Polity.

Sagan, C. (1996) *The Demon-Haunted World: Science as a Candle in the Dark*. New York: Ballantine Books.

Shermer, M. (2016) *Skeptic: Viewing the World with a Rational Eye*. New York: Henry Holt and Company.

Simon, A.E. (1999) *Monsters, Mutants and Missing Links: The Real Science Behind The X-Files*. London: Ebury.

Sismondo, S. (2009) *An Introduction to Science and Technology Studies*. Malden: Blackwell.

White, M. (1998) *The Science of The X-Files*. London: Legend.

White, M. (2005) *A Teaspoon and an Open Mind: The Science of Doctor Who*. London: Allen Lane.

Concluding Remarks
Science in Twenties Doctor Who

LINDY A. ORTHIA AND MARCUS K. HARMES

The essays in this collection traverse *Doctor Who* from 1960s black and white serials up to the most recent broadcasts. Many contributors' ideas were further stimulated by the broadcast of the 2020 season featuring Jodie Whittaker's Doctor, leading to some last minute revisions to essays as we edited the draft volume. The season contained much that was familiar including the return of Captain Jack Harkness, the Judoon, the Master, Gallifrey and the Cybermen. However, the familiarity of such allies, monsters and places has been offset by what reviewers noted as the most extensive "retcon" of the series' history: the revelations in "The Timeless Children" that the origins of the Time Lords and the Doctor were not what we had been led to believe.

Pertinent to this volume, science was intrinsic to these major narrative and thematic developments. This series yields up some level of thematic consistency when read as making stories about science. Some stories foregrounded scientists who have perhaps been neglected by virtue of gender or ethnicity, including Ada Lovelace, the Victorian mathematician ("Spyfall"), and Nikola Tesla, the Serbian-born electrical inventor ("Nikola Tesla's Night of Terror"). In doing so, it counterposed the less familiar with the famous, placing Lovelace alongside the computer pioneer Charles Babbage and Tesla against Thomas Edison. The characterization of both Lovelace and Tesla is complicated. Lovelace (albeit while unconscious, after having her memory wiped) is promised, "Computers start with you." Tesla's characterization is more ambiguous, with his failures asserted and his unfulfilled work contrasted with the business acumen of Thomas Edison. As reviews have since pointed out, at times the characterizations have simplified and even censored; *Doctor Who* asserted the unfairness of Tesla's relative obscurity and heroic failure compared to Edison, with the character

Yaz lamenting, "it's not right." However as his writings show, Tesla's projections as a futurist encompassed not only energy but also purity and he was an enthusiastic eugenicist and advocate for sterilization (Sethi 2016). These aspects of his reputation were likely not relevant to the narrative in "Nikola Tesla's Night of Terror" but constitute knowledge which complicates the broad strokes used in writing his character (Opie 2020).

In other cases, the 2020 episodes have urged viewers to see scientific activity and achievement in unfamiliar places (unfamiliar for a British television audience at least). One such is the humane and caring treatment of mentally ill people in fourteenth-century Aleppo in "Can You Hear Me?" The episode depicts "one of the oldest hospitals in the world" as the Doctor describes it, and characterizes the Islamic physicians running it as, in the Doctor's view, "enlightened." Unusually for *Doctor Who*, this cuts against stereotypical Eurocentric histories of science. Such histories have tended to ignore, marginalize or denigrate Islamic Middle-Eastern and North African scholars' innumerable contributions to science and medicine during the so-called "Islamic Golden Age" of the eighth to fourteenth centuries CE (or indeed at other times in history). In this sense, *Doctor Who* has caught up with more contemporary views of science and medical history (Gorini 2007–2008).

In other ways, the 2020 season's stories have overtly and even stridently foregrounded environmental science. "Praxeus" builds plastic pollution into the plot and the menace, with descriptions of Earth's giant ocean garbage gyres, the presence of microplastics in food, water and bodies, and the consequences of plastic waste for seabirds. Meanwhile, in "Orphan 55," the revelation that a scarred and barren planet is actually Earth, ravaged by global warming, is angrily explained by the Doctor as the consequence of not having listened to "warnings from every scientist alive." She provides a glimmer of hope though, that we can still change it, saying:

> It's one timeline. You want me to tell you the Earth's going to be okay? 'Cause I can't. In your time, humanity's busy arguing over the washing up while the house burns down. Unless people face facts and change, catastrophe is coming. But it's not decided. You know that. The future is not fixed, it depends on billions of decisions, and actions, and people stepping up.

These depictions of scientists and science are markedly different from those seen in earlier seasons of *Doctor Who*. The series has, in every decade of its production, featured many fictitious megalomaniacal, inhumane or heroic scientists, but few real historical scientists have been brought to life onscreen before. The Doctor narrowly missed meeting Leonardo da Vinci in "The Masque of Mandragora" (1976) and "City of Death" (1979), but did meet George Stephenson in "The Mark of the Rani" (1985) and

Albert Einstein in "Time and the Rani" (1987). *Doctor Who*'s previous brief engagements with historical science have also largely reproduced a vision of science centered on Europe, for example through scientist character Giuliano in "The Masque of Mandragora" and Charles Preslin in "The Massacre" (1966) as well as Stephenson and Einstein (see Orthia 2013), rather than introducing viewers to non–Western approaches to science throughout time. While the program has always been political, and has depicted environmental themes numerous times, especially so during the producership of Barry Letts in the mid–1970s (see Harmes' essay in this collection and Orthia 2011), rarely has it been as resolutely furious with environmental destruction as in this series.

This inevitably provokes the question of what point the season is trying to make about science. Asserting the humane treatment of mental illness in Golden Age Syria invites comparison with the treatments offered in contemporary European institutions such as Bedlam (or St Mary of Bethlehem), shown to be a horrific place in "The Shakespeare Code" (2007). The Doctor's introduction to the Aleppo hospital is an overwritten expository piece of dialogue, rendered still clunkier by the fact that she is talking to herself: "Bimaristan. It means sick place. This must be one of the oldest hospitals in the world. Of course, Islamic physicians were known for the enlightened way they treated people with mental health problems." Yet these lines are nonetheless necessary to make the writer's point about the sophisticated nature of Arabic science. Similarly the line "Computers start with you" ripostes the oft-cited patriarchal judgment that Babbage was the "father of computers," instead asserting Lovelace's role in their evolution. It seems that, despite some ongoing criticisms from fans that he is too "woke" or not "woke" enough in his plotting, casting and characterization decisions (Hudson 2020), show-runner Chris Chibnall and the writers and directors he has commissioned have continued to challenge *Doctor Who*'s long history of exclusionary practices in bringing these scientists from the margins to the center of the Whoniverse.

What might viewers have learnt about science and scientists from this season? In the introduction we suggested that early *Doctor Who*, if not exactly educational, might from time to time have imparted occasional scientific lessons, such how condensation forms. Recent stories have also imparted lessons (the dangers of climate change, the unfair neglect of some scientists) in sometimes heavy-handed ways, but have also provided small-scale lessons in science. Thus, viewers of "Praxeus" will have learnt a small nugget of oceanographic knowledge, when the Doctor explains what a gyre is, or when scientist character Suki says she has seen hatchling seabirds regurgitate 200 pieces of plastic, mistaking them for food.

The season's stories have also pinpointed an intriguing interplay between science and science fiction. The juxtaposition of the Lone Cyberman with the elegant interiors of an eighteenth century villa ("The Haunting of Villa Diodati") is a type of juxtaposition of alien modernity with historic European setting that has a long precedence in *Doctor Who*, such as the Daleks in Maxtible's Victorian mansion discussed in the essay by Harmes and Scully, this volume. However, this Cyberman goes further than juxtaposition. The body horror inherent to Cybermen since their first appearance in "The Tenth Planet" (1966) and the grotesque implications of turning bodies into bio-mechanical monsters have been traced to various sources of creative inspiration, including the concerns about transplant surgery and prosthetic surgery held by *Doctor Who* scientific advisor and Cybermen co-creator, Kit Pedler. The body horror of the liminal Cybermen in "The Tenth Planet," as creatures stitched together and with the joins between their parts showing, is also linked to Mary Shelley's 1818 novel *Frankenstein*. But "The Haunting of Villa Diodati" inverts and reverses this adaptive trajectory, instead turning the Cybermen (in particular the partly converted, monstrous lone Cyberman) into the source of Shelley's novel.

While the season has depicted real scientists, it also has its share of fictitious scientist characters. Among these, none is quite as significant as Tecteun from "The Timeless Children." In recounting her story, the Master describes her as "a scientist and explorer," a space-traveling pioneer who finds a lone child on a planet and discovers she can regenerate. Tecteun becomes obsessed with discovering the secret of this ability and eventually does. While much is yet to be discovered about Tecteun's experiments, her story thus far finds resonance with other real or fictitious scientists in two main ways. First, we know she spliced the ability to regenerate into her own genetic code, "to prove herself right." She shares this action with a number of real world scientists who similarly risked their own health to test their theories, such as Nobel Laureate Barry Marshall who in 1984 swallowed a solution of *Helicobacter pylori* bacteria to prove it causes stomach ulcers. In this respect, Tecteun may represent scientists' noble struggle towards discovery. However, she was old and perhaps near death when she undertook the genetic splice, so alternatively she may represent a more sinister, self-serving model of scientist, one that resonates more with villainous scientist archetypes characterized by ambition and inhuman cruelty (Haynes 2003). This is reinforced by the episode's hints that Tecteun may have killed the regenerating child several times, in order to study the regenerative process over and over; actions more reminiscent of the way Josef Mengele joked and gave sweets to the Jewish children before experimenting on them (Lagnado and Dekel 1991). What exactly happened with Tecteun is at the time of writing unknown, but there is a very nasty glimpse of science at the center of "The Timeless Children," which the series will hopefully explore further in the future.

Whatever the coming years hold for *Doctor Who*, we can certainly say this. "The Timeless Children" shows the Whoniverse's foundations were forged in a crucible of science's promises and betrayals, just as *Doctor Who* was created to spark children's interest in science. Science has always been part of *Doctor Who*, from the start, as motivation, justification and inspiration.

REFERENCES

Gorini, R. (2007–2008) Bimaristans and mental health in two different areas of the Medieval Islamic World. *Journal of the International Society for the History of Islamic Medicine* 6–7: 16–20.

Haynes, R. (2003) From alchemy to artificial intelligence: Stereotypes of the scientist in Western literature. *Public Understanding of Science* 12: 243–253.

Hudson, J. (2020, January 9) Too woke? Nope—*Doctor Who* is more offensive than ever, *The Guardian*. www.theguardian.com/tv-and-radio/2020/jan/08/doctor-who-more-offensive-than-ever-jodie-whittaker-pc.

Lagnado, L.M., and Dekel, S.C. (1991) *Children of the Flames: Dr. Josef Mengele and the Untold Story of the Twins of Auschwitz*. New York: William Morrow and co.

Opie, D. (2020, January 19) *Doctor Who*'s portrayal of Nikola Tesla overlooks a MAJOR issue, *Digital Spy*. www.digitalspy.com/tv/a30564587/doctor-who-nikola-tesla-night-of-terror/ (accessed 20 March 2020).

Orthia, L.A. (2011) "Paradise is a little too green for me": Discourses of environmental disaster in *Doctor Who* 1963–2010. *Colloquy* 21: 56–80.

Orthia, L.A. (2013) Savages, science, stagism and the naturalized ascendancy of the Not-We in *Doctor Who*. In: Orthia, L. (ed.) *Doctor Who and Race*. Bristol: Intellect, pp. 269–287.

Sethi, A.K. (2016) *The European Edisons: Volta, Tesla, and Tigerstedt*. London: Springer.

About the Contributors

Lynne **Bowker** is a professor at the School of Translation and Interpretation at the University of Ottawa, Canada, where she teaches and conducts research in the area of translation technologies. She has published widely on the subject of computer-aided and machine translation, including *Computer-Aided Translation Technology* and *Machine Translation and Global Research*. She is also a certified French-English translator.

Vanessa **de Kauwe** is a researcher with expertise in philosophy, ethics, science communication and disability education who submitted her Ph.D. in the latter two fields. She is a practicing disability advocate in Australia, and Ambassador for Disability and Gender Equity internationally. She has published on representations of race and politics in *Doctor Who*, and on other areas in science fiction.

J.J. **Eldridge** is an associate professor and works in the department of Physics at the University of Auckland. Their general research concerns the lives and deaths of stars, from those in our own galaxy to those in galaxies at the edge of the observable universe.

Mark **Erickson** is reader in sociology and the director of postgraduate studies at the University of Brighton, UK. He is a sociologist with research interests in sociology of science and technology, cultural studies of science and technology, social theory, and social research methods and methodology. He is the author of *Science, Culture and Society*.

Ross **Garner** is a lecturer in media and cultural studies in the School of Journalism, Media and Culture at Cardiff University. He has published multiple articles relating to *Doctor Who*, including in *Popular Communication* and *Tourist Studies*. His research is on the meanings and significance of different mediations of the Mesozoic within popular culture alongside his interests in media tourism, mediated nostalgia, television and transmedia studies.

Mark **Halley** is an assistant professor of American sign language/English interpreting at the University of North Florida. He conducted his doctoral research in the Department of Interpretation and Translation at Gallaudet University in Washington, D.C., where he studied the role of interpreters in the 1988 Deaf President Now protest. He has also been an interpreter in private practice since 2011.

About the Contributors

Marcus K. **Harmes** is an associate professor at the University of Southern Queensland and is the author of *Roger Delgado: I Am Usually Referred to as the Master* (2018), *Doctor Who and the Art of Adaptation* (2013) and other studies on the science-fiction television program. His other areas of research include the cultural history of education (which also includes *Doctor Who*) and legal history.

Kristine **Larsen** is a professor of astronomy at Central Connecticut State University. Her teaching and research focus on the intersections between science and society, including the history of women in science and depictions of science in popular media (especially the works of J.R.R. Tolkien). She is the author of *Particle Panic!*, *The Women Who Popularized Geology in the 19th Century*, *Cosmology 101*, and *Stephen Hawking: A Biography*.

Catriona **Mills** is the content manager of AustLit. She holds degrees from Macquarie University and the University of Queensland, with a specialty in nineteenth-century periodical literature. She has published on adaptations of penny-weekly serials to the English suburban stage, authorship attribution in Australian nineteenth-century periodicals, steampunk's problematic relationship to empire, and *Doctor Who*.

Lindy A. **Orthia** is a science communication researcher at the Australian National University. Her primary research interest is in the politics of knowledge, especially intersections between science and race, gender, sexuality and class. Her doctoral research examined the social, political, ideological, cultural and economic aspects of science in *Doctor Who*, and she has since published extensively on this. She edited the critically acclaimed volume *Doctor Who and Race*.

Natalie **Ring** is a postdoctoral research fellow at the Roslin Institute in Edinburgh, Scotland, where her research focuses on the genetics of infectious diseases. She has a postgraduate diploma in science communication from the University of Edinburgh, and a Ph.D. in microbial genomics from the University of Bath.

Richard **Scully** is an associate professor in modern history at the University of New England. His research interests are in the history of cartoons and satirical art, and Anglo-German relations (1860–1914). His work includes *Eminent Victorian Cartoonists* (3 volumes, 2020), and a chapter for Lindy A. Orthia's *Doctor Who and Race* (2013).

Tonguç İbrahim **Sezen** is a lecturer of transmedia production at Teesside University, School of Computing & Digital Technologies. He has been a Fulbright scholar at Georgia Institute of Technology, School of Literature, Media, and Communication and a research fellow for the Rheijnland.Xperiences innovation project. His research interests include game design, interactive storytelling, and toy studies. He is one of the editors of *Interactive Digital Narrative* (2015).

Mike **Stack** completed a Ph.D. in psychosocial studies at Birkbeck, University of London, with a thesis entitled "Queer Who: Doctor Who Fandom, Gay Male Subculture and Transitional Space." He is grateful to Dr. Annette-Carina van der Zaag, for whom he provided teaching assistance on the second-year module Sexuality, for her expertise and suggestions on the science of sex and gender.

Elizabeth R. **Stanway** is an associate professor in astronomy and astrophysics at the University of Warwick in the UK. Her work involves the investigation and modeling of stellar populations in galaxies in the distant universe. She is also a member of Warwick's Centre for Exoplanets and Habitability and has published studies of the potential habitability of galaxies, investigating how supernovae, gamma ray bursts and other explosive events affects this over cosmic time.

Index

Note: The Doctor's companions are indexed by first name except Ace (under "A") and Brigadier Lethbridge-Stewart (under "B"). Numbers in **bold italics** indicate pages with illustrations.

Abbott, Edwin Abbott 13, 157, 159, 161–162
Ace 8, 98
Adams, Douglas 6, 62
Adric 7, 160
"The Age of Steel" see "Rise of the Cybermen"
alchemy 127, 142, 143, 145, 150, 151, 154
"The Almost People" see "The Rebel Flesh"
"The Ambassadors of Death" 42, 64
Amy Pond 98, 129, 180
"Amy's Choice" 19
Apollo 23 36, 39
"Arachnids in the UK" 115, 120
archaeology 7, 9, 13, 134, 152, 157, 158, 162–168, 171, 195, 199
"The Ark in Space" 112
"The Armageddon Factor" 81
"Army of Ghosts/Doomsday" 160
artificial intelligence 10, 68, 70, 212
"Ascension of the Cybermen" see "The Timeless Children"
astronomy and astrophysics 8, 12, 18–30, 39–41, 56, 117, 151, 216
"Asylum of the Daleks" 105
atomic energy see nuclear technology
"Attack of the Cybermen" 39
Automatic Language Processing Advisory Committee (ALPAC) 67
"The Aztecs" 4, 164

"Bad Wolf/The Parting of the Ways" 39, 79, 82
Baker, Colin and Sixth Doctor 1, 7, 15
Baker, Tom and Fourth Doctor 1, 7, 8, 42, 76*n*1, 79, 107*n*2, 110, 116, 118, 120, 131, 179, 180, 184, 210
Barbara Wright 4, 5, 63, 98, 169, 170
"The Battle of Ranskoor Av Kolos" 120
"Battlefield" 164

"Before the Flood" see "Under the Lake"
Bidmead, Christopher H. 6, 110
"The Big Bang" see "The Pandorica Opens"
Bill Potts 7, 83, 114, 137
biologism 99, 100, 102, 103
"Black Orchid" 168
Blood Heat 187
"Bloodtide" 187
brain 70, 90–91, 99, 102–103, 132, 133, 135
"The Brain of Morbius" 135
Brigadier Lethbridge-Stewart 7, 8, 64

"Can You Hear Me?" 120, 222
Capaldi, Peter and Twelfth Doctor 1, 9, 13, 79, 81, 82, 83, 88, 90, 97, 98, 136, 160, 171, 178, 179, 182, 185
"Carnival of Monsters" 168
Carroll, Lewis 144, 159
"Castrovalva" 160
"The Celestial Toymaker" 19, 160
cells (biological) 12, 38, 83–87, 89, 90, 92*n*2
"The Chase" 24, 169
Chibnall, Chris 79, 95, 223
"The Christmas Invasion" 71, 81, 84
chromosomes 88, 92*n*3, 98, 99, 105
"City of Death" 168, 222
Clara Oswald 41, 74, 81, 105, 136, 160, 161
class and classism 7, 8, 104, 105, 116, 119, 123, 153, 154, 182
"The Claws of Axos" 54
climate and climate change 26, 117, 120, 222, 223
"Closing Time" 114, 134
coal 12, 49, 53, 55, 57–58, 120, 132
"Cold Blood" see "The Hungry Earth"
"Cold War" 74, 165
colonialism and colonization 10, 12, 13, 19, 25–26, 35, 39, 153, 158, 162–165, 167–170
"Colony in Space" 7, 26

231

"The Crimson Horror" 153
Cuban Missile Crisis 49, 50, 57
"The Curse of Peladon" 22, 106, 164
Cushing, Peter 148

"The Dæmons" 48, 53, 118
"Dalek" 100, 168
"The Dalek Invasion of Earth" 4, 27, 51, 60
"The Daleks" 50, 59, 60, 63, 193
"Daleks in Manhattan/Evolution of the Daleks" 100
"The Daleks' Master Plan" 51
Danny Pink 7, 8
"Dark Water/Death in Heaven" 136, 137, 138
Davies, Russell T 79
Davis, Gerry 78
Davison, Peter and Fifth Doctor 1, 7, 112, 118, 119, 177, 185
"The Day of the Doctor" 79, 80, 113, 131
"Day of the Moon" *see* "The Impossible Astronaut"
"The Deadly Assassin" 90
"Death to the Daleks" 54
"Deep Breath" 81, 174, 178, 179, 181, 182
deep time 14, 163, 167, 170, 173, 183, 184
"Demons of the Punjab" 97, 104, 119
"Destiny of the Daleks" 81, 107*n*2
dimensionality 13, 145, 157, 158–162, 164, 171
dinosaurs 14, 68, 173–188, 201, 202
"Dinosaurs on a Spaceship" 38, 68, 174, 178, 180, 181, 184
disability 119
"The Doctor Falls" *see* "World Enough and Time"
"The Doctor's Wife" 83, 160
"The Dominators" 168
Donna Noble 65
"Doomsday" *see* "Army of Ghosts"
Dust of Ages 36, 37, 41

Earth-like planet 18, 19–20, 21, 23, 25, 30*n*1, 43
"The Eaters of Light" 137
Eccleston, Christopher and Ninth Doctor 1, 9, 79, 80, 82
"The Edge of Destruction" 50
Eighth Doctor *see* McGann, Paul
Eleventh Doctor *see* Smith, Matt
"The Eleventh Hour" 129
empire *see* colonialism and colonization
"Empress of Mars" **42**, 106, 165
"The End of the World" 66
"The End of Time" 136
"The Enemy of the World" 52, 112
energy crisis 57–58
environmentalism 5, 10, 14, 26, 55, 72, 121, 175, 185, 209, 222, 223
"The Evil of the Daleks" 13, 53, 142–154
exoplanet 18–19, 23, 24
"Extremis" 137, 168
The Eye of the Giant 71

"The Faceless Ones" 59
fans 11, 14, 24, 41, 95, 110, 136, 139, 173, 192, 193, 197, 198–199, 202, 205–206, 212, 217, 223
Faraday, Michael 13, 142, 145, 149, 150, 151, 152, 213, 214, 216
FASA Doctor Who roleplaying game 200, 201
Fifth Doctor *see* Davison, Peter
"The Fires of Pompeii" 65
First Doctor *see* Hartnell, William
"The Five Doctors" 80, 82
"Flatline" 13, 157, 158, 160–162, 170–171
"Flesh and Stone" *see* "The Time of Angels"
"Forest of the Dead" *see* "Silence in the Library"
"Four to Doomsday" 64, 70
fourth dimension *see* dimensionality
Fourth Doctor *see* Baker, Tom
"Frontier in Space" 34, 39
"Fugitive of the Judoon" 80, 94, 107, 110, 122
"Full Circle" 116

"Galaxy Four" 7
gender 4, 9, 13, 15, 35, 40, 73, 83, 94–108, 110–111, 119, 123, 127–139, 153, 209, 211, 216–217, 218, 221
genetics 9, 10, 59, 82, 83, 84, 86, 87–90, 92*n*3, 99–100, 102, 105, 106, 107, 120, 132, 133, 212, 215–216, 224
"Ghost Light" 152, 153, 168
"The Ghost Monument" 119, 120
"The Girl in the Fireplace" 64
gisting (language) 72
"A Good Man Goes to War" 65, 82
Grace Holloway 7
Graham O'Brien 99, 104, 119, 122, 123
"The Greatest Show in the Galaxy" 168
"The Green Death" 8, 58
"Gridlock" 24

Haggard, H. Rider 158, 164, 167, 192
"The Hand of Fear" 164
Harry Sullivan 7
Hartnell, William and First Doctor 1, 2, 3, 6, 7, 53, 78, 91, 112, 118, 119, 157, 171
"The Haunting of Villa Diodati" 63, 122, 224
Heath, Edward 57
"Hell Bent" 83
Hinchcliffe, Philip 79
Hinton, C.H. 145, 147, 158–159
"The Horns of Nimon" 55
"Horror of Fang Rock" 9
"The Hungry Earth/Cold Blood" 38, 130, 165, 175
Hurt, John and War Doctor 1, 2, 79, 80, 82

Ian Chesterton 4, 5, 7, 63, 98, 169
"The Ice Warriors" 22, 164
"Image of the Fendahl" 54, 164
Imperial Moon 35, 71, 72
imperialism *see* colonialism and colonization

"The Impossible Astronaut/Day of the Moon" 35, *42*
"The Impossible Planet/The Satan Pit" 27, 70
Indigenous knowledges (on Earth) 64, 113, 121
"Inferno" 52, 53, 160
interpretation (language) 12, 65–66, 74
"The Invasion" 39
"Invasion of the Dinosaurs" 58, 173–181, 185
"The Invasion of Time" 160
"The Invisible Enemy" 21
"It Takes You Away" 123

"The Jabari Countdown" 9
Jamie McCrimmon 15, 118, 165
Jo Grant 8, 106, 118
"Journey to the Centre of the TARDIS" 160
"Journey's End" *see* "The Stolen Earth"
"Judoon in Chains" 73

Kelvin, Lord (William Thompson) 145–146, 152
"The Keys of Marinus" 168
"Kill the Moon" 36–38, 41, *42*, 43, 45
"Kinda" 101
Kovarian *see* Madame Kovarian
"The Krotons" 24, 54, 116

Lambert, Verity 4, 5, 116
"Last of the Time Lords" *see* "The Sound of Drums"
Leela 8, 9, 118
"Let's Kill Hitler" 39
Letts, Barry 6, 79, 223
LGBTQIA people and topics 9, 15, 95–96, 97, 100–101, 103, 104, 106, 107, 107n1, 108n4, 110, 123
licensing 199
"The Lie of the Land" 137
Liz Shaw 7, 8, 64
localization (language) 67
"The Lodger" 97
"The Long Game" 106, 116
Lovelace, Ada 120, 221, 223

machine bias 73, 75
machine translation 12, 62–76
Madame Kovarian 13, 128, 131, 134–135, 138, 139
"The Magician's Apprentice/The Witch's Familiar" 74, 105, 136
"Marco Polo" 5, 164
"The Mark of the Rani" 131, 132, 222
Mars 12, 18, 21–23, 25, 34, 41–45, *42*, 152
Martha Jones 7, 8
Martin, Jo 1, 2, 79, 80, 107, 110
"The Masque of Mandragora" 55, 63, 66, 222, 223
"The Massacre" 223
The Master and Missy 9, 13, 48, 68, 80, 82, 83, 90, 95, 102, 112, 128, 131, 132, 133, 136–139, 200, 221, 224

Maxwell, James Clerk 13, 142, 143, 144, 145, 149–151, 152, 214, 216
McCoy, Sylvester and Seventh Doctor 1, 7, 98, 112, 114
McGann, Paul and Eighth Doctor 1, 80, 82, 89
Mel Bush 7, 98
melodrama 14, 175, 180–182, 183, 185, 186, 187
memory 68, 71, 80, 81, 90–91, 167, 169, 197, 221
mesmerism 142, 143, 145, 150, 151
Mickey Smith 8
"The Mind of Evil" 64
"The Mind Robber" 8, 160
Missy *see* The Master
Moffat, Steven 79, 139, 146
"The Monster of Peladon" 164–165
monstrosity 131, 133, 134, 135, 136, 174, 175, 179, 180, 182, 186, 224
The Moon 6, 12, 33–45
"The Moonbase" 34, 37, 39, 40, 51, 63, 112
museums 11, 13, 153, 157, 158, 162, 164, 167–170, 171, 177, 184
"The Mutants" 26
"The Myth Makers" 164

Nathan-Turner, John 132
Nation, Terry 193, 194, 197, 202
New Space Age 34, 44–45
Newman, Sydney 3, 5, 6
"The Night of the Doctor" 80, 82, 89
"Nightmare of Eden" 168
"Nikola Tesla's Night of Terror" 221–222
"1963: The Space Race" 36
Ninth Doctor *see* Eccleston, Christopher
novum 195–197, 201
nuclear technology 10, 12, 20, 39, 49, 50, 52, 55–57, 59, 60, 135, 190, 194
Nyssa 7, 8

oil (fossil fuel) 12, 49, 55, 57–58
Only Human 73
"Orphan 55" 65, 120, 222
Osgood, Petronella 130, 131, 136
The Outer Space Treaty 35, 36

"The Pandorica Opens/The Big Bang" 168–169
paratexts 11, 14, 40–41, 187, 190–202, 205–206, 211–218
"The Parting of the Ways" *see* "Bad Wolf"
Pedler, Kit 5, 40, 224
Peri Brown 7, 8, 15, 132, 164
Pertwee, Jon and Third Doctor 1, 7, 8, 26, 42, 52, 54, 79, 80, 116, 118, 120, 176, 179, 180, 185, 210
"The Pilot" 114
"The Pirate Planet" 6, 27, 53, 55
"Planet of Evil" 210
"Planet of Fire" 8, 164
"Planet of Giants" 5, 51
"Planet of the Daleks" 195

234 Index

"The Poison Sky" *see* "The Sontaran Stratagem"
popular science 14, 20, 41, 43, 55, 56, 60, 63, 74, 103, 154, 158, 162, 163, 167, 176, 177, 183, 187, 188n1, 205, 206–218
posthumanism 105, 106
"The Power of the Daleks" 50, 78
"Praxeus" 90, 111, 120, 222, 223
public perceptions of science *see* popular science
"Pyramids of Mars" 42, 152, 164, 165

queer (people) *see* LGBTQIA people and topics

race and racism 7, 8, 35, 58, 64, 73, 104, 107, 116, 121–122, 153, 158, 163–164, 166, 171, 184–185, 222
radiation 21, 52, 59
The Rani 9, 13, 68, 128, 131–133, 135, 138, 139
"The Rebel Flesh/The Almost People" 134
"Red Dawn" 45n5
reference books 192, 197, 198, 199, 201, 202
regeneration 9, 12, 13, 78–91, 95, 99, 100, 101, 102, 104, 107, 134, 136, 139, 170, 224
"The Reign of Terror" 63
religion 10, 119, 122, 134, 135, 138, 143, 151, 153, 209, 210, 218, 222
"Resolution" **20**, 117, 122, 164
retroactive continuity 198–199
"Revenge of the Cybermen" 27
"The Ribos Operation" 64
"Rise of the Cybermen/The Age of Steel" 105, 160
River Song 7, 65, 82, 131, 134, 135, 138, 164
"The Robots of Death" 118
Romana 6, 8, 55, 81, 107n2, 112
"The Romans" 164
Rory Williams 7, 180
"Rosa" 121
Rose Tyler 9, 82, 100, 168, 217
Ryan Sinclair 99, 104, 119, 123

Salisbury, Lord (Robert Gascoyne-Cecil) 145, 146, 152
The Sarah Jane Adventures 9
Sarah Jane Smith 9, 63, 66, 70
"The Satan Pit" *see* "The Impossible Planet"
"The Savages" 54
Scales of Injustice 187
science history 11, 21, 23, 91, 111, 113, 143, 145–146, 148–154, 158, 162–163, 168, 170, 183, 209–210, 211, 213–214, 218, 222, 223
scientific gaze 14, 175–177
scientific romances 13, 33, 147, 157, 158–160, 162, 165, 166, 167, 192
scientism 9, 118, 121, 209, 210, 215, 218
scientists 3, 4, 6–10, 12, 13, 25, 40, 49, 52, 55, 56, 58, 62, 75, 102, 103, 108n5, 110–120, 121, 123, 127–139, 142–154, 158, 162, 165, 167, 175,

183, 196, 207, 208, 209, 210, 213, 214, 215, 216, 217, 218n1, 221, 222–223, 224
"The Sea Devils" 23, 165, 173, 175
Second Doctor *see* Troughton, Patrick
"The Seeds of Death" 22, 26, 34, 37, 39, 40, 164, 168
"The Seeds of Doom" 168
Seventh Doctor *see* McCoy, Sylvester
sex 13, 83, 87, 88–89, 94–107, 217
sexism 7, 13, 73, 95, 103, 110–111, 123, 127, 128, 129–139, 154, 184–185, 217, 218
"The Shakespeare Code" 223
"Silence in the Library/Forest of the Dead" 164
"The Silurian Candidate" 187
The Silurian Gift 187
"The Silurians" 6, 8, 23, 38, 52, 53, 165, 173–180, 184
"Silver Nemesis" 39
Sixth Doctor *see* Baker, Colin
"Sleep No More" 101
"Smith and Jones" 36, **42**, 73
Smith, Matt and Eleventh Doctor 1, 13, 79, 80, 81, 82, 83, 97, 98, 114, 129, 123, 139, 180, 185, 212, 216
"Snakedance" 168
Snow, C.P. 56
solar system 18, 20–23, 24, 27, 29, 42, 50–51, 208
"The Sontaran Stratagem/The Poison Sky" 26
"The Sound of Drums/Last of the Time Lords" 82, 106
Souvenir Press 193–197, 198
"The Space Museum" 13, 158, 168, 169, 170, 171
"Spearhead from Space" 48, 71, 113
spectacle 45, 144, 175–179, 180, 182, 183, 186
speech recognition 65
"Spyfall" 90, 221
"State of Decay" 53, 66, 114
static electricity 50, 53, 142, 145, 146, 149, 150, 193
stereotypes 7, 13, 83, 98, 103, 104, 108n4, 108n5, 123, 127–131, 132, 133, 135, 138, 139, 218, 222
stewardship 13, 111, 121–123
Stewart, Kate 131, 136
"The Stolen Earth/Journey's End" 80, 82, 84
"The Stones of Blood" 9, 53, 164
"The Sun Makers" 21
Susan Foreman 4, 5, 63

"The Talons of Weng-Chiang" 54, 64, 152, 165
Tegan Jovanka 64, 177
Tennant, David and Tenth Doctor 1, 9, 79, 80, 81, 82, 84, 112, 116, 119, 123, 136
Tenth Doctor *see* Tennant, David
"The Tenth Planet" 34, 53, 78, 224
"Terminus" 8
terraforming 12, 23, 25–26
"Terror of the Autons" 8, 166
"Terror of the Zygons" 66, 70, 174, 177, 179, 180

Tesla, Nikola 120, 221–222
Third Doctor *see* Pertwee, Jon
Thirteenth Doctor *see* Whittaker, Jodie
"A Thousand Tiny Wings" 73
"The Three Doctors" 54, 57, 80
"Time and the Rani" 131, 133, 223
Time Lords' gift of translation 64, 66, 68, 75
"The Time Meddler" 59, 195
"The Time Monster" 48, 71, 164
"The Time of Angels/Flesh and Stone" 164
"The Time of the Doctor" 82, 134
"The Time Warrior" 116
"The Timeless Children" 78, 80, 82, 90, 94, 106, 107, 110, 221, 224–225
"The Tomb of the Cybermen" 13, 157, 158, 165–167, 170, 171
"Tooth and Claw" 38
Torchwood 8, 187
trans (people) *see* LGBTQIA people and topics
translation circuit 12, 62–75, 117
"The Trial of a Time Lord" 24
Troughton, Patrick and Second Doctor 1, 7, 15, 51, 78, 110, 112, 118, 119
"The Tsuranga Conundrum" 101, 110, 120
Twelfth Doctor *see* Capaldi, Peter
"Twice Upon a Time" 80, 82, 83
"The Two Doctors" 15, 64, 80

"Under the Lake/Before the Flood" 69
"The Underwater Menace" 164
"An Unearthly Child" 5, *20*, 49, 63, 78, 119, 173
"Utopia" 136

Vicki 169
Victoria Waterfield 144, 150, 157, 165

Victorian era 13, 35, 112, 142–154, 157–171, 221, 224
"Victory of the Daleks" 39
villains 13, 57, 68, 127, 128, 131, 135, 136, 138, 139, 161, 165, 182, 190, 224
"Vincent and the Doctor" 168, 169

War Doctor *see* Hurt, John
"The War Games" 63, 78, 81
"Warriors' Gate" 160
"Warriors of the Deep" 165, 174, 177, 179
"The Waters of Mars" 23, *42*, 44, 119
Weaver's memorandum 66–67
"The Web of Fear" 168
"The Web Planet" 18
"The Wedding of River Song" 131, 135
Wells, H.G. 22, 26, 33, 58, 127, 144, 145, 147, 152, 157, 159, 170
"The Wheel in Space" 118
Whitaker, David 149, 151, 152, 193
Whittaker, Jodie and Thirteenth Doctor 1, 7, 13, 14, 79, 80, 81, 83, 88, 90, 94–95, 98, 99, 102, 104, 107, 110–112, 116, 117, 119–123, 139, 185, 221
Wilson, Harold 56
"The Witchfinders" 97, 122
"The Witch's Familiar" *see* "The Magician's Apprentice"
"The Woman Who Fell to Earth" 81, 97, 100, 104, 116, 122
"World Enough and Time/The Doctor Falls" 83, 137

Yaz Khan 104, 119, 222

Zoe Heriot 7, 8, 40, 118

www.ingramcontent.com/pod-product-compliance
Ingram Content Group UK Ltd.
Pitfield, Milton Keynes, MK11 3LW, UK
UKHW041943140426
5217IPUK00014B/626